# Monographs in Electrical and Electronic Engineering 21

Series editors: P. Hammond, T. J. E. Miller, and S. Yamamura

# Monographs in Electrical and Electronic Engineering

10. *The theory of linear induction machinery* (1980)
    Michel Poloujadoff
12. *Energy methods in electromagnetism* (1981) P. Hammond
15. *Superconducting rotating electrical machines* (1983) J. R. Bumby
16. *Stepping motors and their microprocessor controls* (1984) T. Kenjo
18. *Permanent-magnet and brushless d.c. motors* (1985) T. Kenjo and
    S. Nagamori
19. *Metal–semiconductor contacts. Second edition* (1988) E. H. Rhoderick and
    R. H. Williams
20. *Introduction to power electronics* (1988) Eiichi Ohno
21. *Brushless permanent-magnet and reluctance motor drives* (1989)
    T. J. E. Miller
22. *Vector control of a.c. machines* (1990) Peter Vas
23. *Brushless servomotors: fundamentals and applications* (1990)
    Y. Dote and S. Kinoshita
24. *Semiconductor devices, circuits, and systems* (1991)
    Albrecht Möschwitzer
25. *Electrical machines and drives: a space-vector theory approach*
    (1992) Peter Vas
26. *Spiral vector theory of a.c. circuits and machines* (1992)
    Sakae Yamamura
27. *Parameter estimation, condition monitoring, and diagnosis of electrical
    machines* (1993) Peter Vas
28. *An introduction to ultrasonic motors* (1993) S. Sashida and T. Kenjo
29. *Ultrasonic motors: theory and applications* (1993) S. Ueha and
    Y. Tomikawa
30. *Linear induction drives* (1993) J. F. Gieras
31. *Switched reluctance motors and their control* (1993) T. J. E. Miller
32. *Numerical modelling of eddy currents* (1993) Andrzej Krawczyk
    and John A. Tegopoulos

# Brushless Permanent-Magnet and Reluctance Motor Drives

T. J. E. Miller

GEC Titular Professor in Power Electronics
University of Glasgow

CLARENDON PRESS · OXFORD

Oxford University Press, Walton Street, Oxford OX2 6DP
Oxford New York Toronto
Delhi Bombay Calcutta Madras Karachi
Kuala Lumpur Singapore Hong Kong Tokyo
Nairobi Dar es Salaam Cape Town
Melbourne Auckland Madrid
and associated companies in
Berlin Ibadan

Oxford is a trade mark of Oxford University Press

Published in the United States
by Oxford University Press Inc., New York

© T. J. E. Miller, 1989
Reprinted with corrections 1993

All rights reserved. No part of this publication may be reproduced,
stored in a retrieval system, or transmitted, in any form or by any means,
electronic, mechanical, photocopying, recording, or otherwise, without
the prior permission of Oxford University Press

British Library Cataloguing in Publication Data
Miller, T. J. E. (Timothy John Eastham), 1947–
Brushless permanent-magnet and reluctance
motor drives.
1. Direct current electric motors
I. Title  II. Series
621.46'2
ISBN 0-19-859369-4

Library of Congress Cataloging in Publication Data
Miller, T. J. E. (Timothy John Eastham), 1947–
Brushless permanent-magnet and reluctance
motor drives.
(Monographs in electrical and electronic engineering; 21)
Bibliography: p.  Includes index.
1. Electric motors, Direct current.  2. Electric
motors, Brushless.  3. Reluctance motors.  I. Title.
II. Series
TK2681.M55  1989  621.46'2  88-23173
ISBN 0-19-859369-4

Printed in Great Britain by
Butler and Tanner Ltd, Frome, Somerset

# Preface

The impulse to write this book was most recently inspired by the publication of Professor Kenjo's books in the same series, the idea being to extend the coverage and provide more detail on synchronous brushless motors and the switched reluctance motor. However, the basic idea of a book in this area goes back several years to a period of particularly interesting developments under the Motor Technology Program at the Corporate Research and Development Center of General Electric in Schenectady, New York. This programme was coupled with exciting developments in semiconductors and power electronics (Baliga 1987), as well as with rapid changes in the technology of motor drives originating in all parts of the developed world. While I was privileged to participate in this programme I also had the benefit of having worked under Professor Peter Lawrenson at Leeds University. The pressures of business prevented any writing until I accepted my present post at Glasgow University, which is supported by GEC, UK. The Scottish Power Electronics and Electric Drives Consortium (SPEED), established in 1986 and modelled on the Wisconsin Electric Machines and Power Electronics Consortium, has provided an environment for further analysis and experimentation, as well as new results and perspectives, and an appreciation of the need for a text in this area. In writing it, I claim no credit for the original inventions or for anything more than a small part in their subsequent development; the book is merely intended as a reasonably organized account of the fundamental principles.

It is hoped that this presentation of the theory of operation of brushless d.c. drives will help engineers to appreciate their potential and apply them more widely, taking advantage of remarkable recent developments in permanent-magnet materials, power semiconductors, electronic control, and motor design (including CAD). The objective is not to 'sell' particular technologies or teach design, but to lay out the basic principles, and it is hoped that this will raise the general credibility and acceptance of new technology that many engineers have striven to establish. It is also hoped that the sections on permanent magnets and magnetic circuits will assist in the exploitation of new PM materials with outstanding properties greatly improved from those of only a few years ago.

It is humbling to realize how much scope for innovation remains in the field of motors and drives, even a century and a half after Faraday. Yet in the academic world the subject of motor design and power engineering more generally has fallen into such decline that the demand for power engineers exceeds the supply, and motor designers are scarce. Some of the present material was developed for courses at Drives, Motors, and Controls Conference and at the University of Wisconsin, and this book is addressed to some of industry's educational needs. Examples and problems are included, many of which were developed as tutorial material for (and by!) students at Glasgow University.

The approach taken is essentially academic: theory and calculation predominate, and the really difficult questions of comparisons between different drives, and the design of particular ones, are treated only lightly. It is hoped, however, that most of the basic theory of modern brushless drives will be found here. The treatment of magnetic saturation is given less attention than in classical works on electric motors: in the design of brushless motors, it is important to grasp the first principles, which can be understood, in the main, from linear theory. The widespread availability of finite-element analysis, and its ever-improving capability, make the problems of saturation much more tractable and relieve the need for a more complex analytical approach, which would be exceedingly complex before it could be really useful.

If nothing else, a study of brushless motor drives will lead to a further appreciation of the extraordinary properties of conventional motors, particularly the d.c. commutator motor and the a.c. induction motor, and will throw a little light on the achievements of our forebears. The arrival of silicon power electronics has reopened all the fundamental questions, and added a new dimension to the equation that has for so long been dominated by copper and iron.

*Glasgow*  T. J. E. M.
April 1988

# Acknowledgements

Many engineers have contributed to this book through the lessons they have taught me. Most of their original work is in print and referenced throughout the book, but particular acknowledgement is made to those with whom I have worked, including Dr Eike Richter and Dr Edward P. Cornell of GE, together with many others, of whom I would particularly like to mention G. B. Kliman, T. M. Jahns, T. W. Neumann, D. M. Erdman, H. B. Harms, F. L. Forbes, and V. B. Honsinger. Also to be particularly recognized is the work of Professor P. J. Lawrenson and Dr Michael Stephenson and their colleagues at the University of Leeds, where much of the European work on reluctance machines (both synchronous and switched) originated; and of Professor M. R. Harris of the University of Newcastle upon Tyne. Several new ideas and experimental results were contributed by Peter Bower under the Glasgow University SPEED Programme, and acknowledgement is made to Anderson Strathclyde plc, Emerson Electric, Lucas, National Semiconductor, Pacific Scientific, Simmonds Precision, Hoover, Reliance Electric, Smith and Nephew, and GE under this programme. Acknowledgement is also made to Professor J. Lamb and the University Court of the University of Glasgow, and to Professor P. Hammond for his encouragement.

To my family and friends

# Contents

GLOSSARY OF SYMBOLS — xiii

## 1 INTRODUCTION — 1

1.1 Motion control systems — 1
1.2 Why adjustable speed? — 1
    1.2.1 Large versus small drives — 4
1.3 Structure of drive systems — 5
1.4 New technology — 7
    1.4.1 Digital electronics — 7
    1.4.2 Power integrated circuits — 8
    1.4.3 Power semiconductor devices — 9
    1.4.4 New magnetic materials — 9
    1.4.5 CAD and numerical analysis in design — 9
    1.4.6 Other contributing technologies — 10
1.5 Which motor? — 11
    1.5.1 Evolution of motors — 11
    1.5.2 The d.c. commutator motor — 13
    1.5.3 The PM d.c. commutator motor — 13
    1.5.4 The induction motor drive — 16
    1.5.5 The brushless d.c. PM motor — 17
    1.5.6 The brushless PM a.c. synchronous motor — 17

## 2 PRINCIPLES OF SIZING, GEARING, AND TORQUE PRODUCTION — 20

2.1 Sizing an electric motor — 20
    2.1.1 Airgap shear stress — 20
    2.1.2 Torque per unit stator volume — 23
2.2 Choice of gear ratio in geared drives — 24
    2.2.1 Simple acceleration of pure inertia load — 25
    2.2.2 Acceleration of inertia with fixed load torque — 26
    2.2.3 Peak/continuous torque ratio of motor — 26
    2.2.4 General speed and position profiles — 27

|     |     |     |
| --- | --- | --- |
| 2.3 | Basic principles of torque production | 28 |
|     | 2.3.1 Production of smooth torque | 29 |
| Problems for Chapter 2 | | 32 |

## 3 PERMANENT-MAGNET MATERIALS AND CIRCUITS     34

3.1 Permanent-magnet materials and characteristics     34

3.2 $B$-$H$ loop and demagnetization characteristics     35

3.3 Temperature effects: reversible and irreversible losses     41
    3.3.1   High-temperature effects     41
    3.3.2   Reversible losses     41
    3.3.3   Irreversible losses recoverable by remagnetization     43

3.4 Mechanical properties, handling, and magnetization     44

3.5 Application of permanent magnets in motors     46
    3.5.1   Power density     46
    3.5.2   Operating temperature range     47
    3.5.3   Severity of operational duty     47

Problems for Chapter 3     50

## 4 SQUAREWAVE PERMANENT-MAGNET BRUSHLESS MOTOR DRIVES     54

4.1 Why brushless d.c.?     54

4.2 Magnetic circuit analysis on open-circuit     58

4.3 Squarewave brushless motor: torque and e.m.f. equations     63

4.4 Torque/speed characteristic: performance and efficiency     66

4.5 Alternative formulations for torque and e.m.f.     68

4.6 Motors with 120° and 180° magnet arcs: commutation     70

4.7 Squarewave motor: winding inductances and armature reaction     76

4.8 Controllers     80

4.9 Computer simulation     83

Problems for Chapter 4     85

## CONTENTS

**5 SINEWAVE PERMANENT-MAGNET BRUSHLESS MOTOR DRIVES** — 88

    5.1 Ideal sinewave motor: torque, e.m.f., and reactance — 89
        5.1.1 Torque — 89
        5.1.2 E.m.f. — 92
        5.1.3 Inductance of phase winding — 94
        5.1.4 Synchronous reactance — 96

    5.2 Sinewave motor with practical windings — 96

    5.3 Phasor diagram — 100

    5.4 Sinewave motor: circle diagram and torque/speed characteristic — 103

    5.5 Torque per ampere and kVA/kW of squarewave and sinewave motors — 109

    5.6 Permanent magnet versus electromagnetic excitation — 112

    5.7 Slotless motors — 115

    5.8 Ripple torque in sinewave motors — 116

    Problems for Chapter 5 — 117

**6 ALTERNATING-CURRENT DRIVES WITH PM AND SYNCHRONOUS-RELUCTANCE HYBRID MOTORS** — 118

    6.1 Rotors — 118

    6.2 A.c. windings and inductances — 122
        6.2.1 Open-circuit e.m.f. — 122
        6.2.2 Synchronous reactance (d-axis) — 128
        6.2.3 Synchronous reactance (q-axis) — 131
        6.2.4 Magnet flux-density and operating point — 134

    6.3 Steady-state phasor diagram — 135
        6.3.1 Converter volt-ampere requirements — 144

    6.4 Circle diagram and torque/speed characteristic — 145

    6.5 Cage-type motors — 147

    Problems for Chapter 6 — 148

**7 SWITCHED RELUCTANCE DRIVES** — 149

    7.1 The switched reluctance motor — 149

|     |     |
| --- | --- |
| 7.2 Poles, phases, and windings | 156 |
| 7.3 Static torque production | 158 |
|     7.3.1 Energy conversion loop | 164 |
| 7.4 Partition of energy and the effects of saturation | 168 |
| 7.5 Dynamic torque production | 172 |
| 7.6 Converter circuits | 173 |
| 7.7 Control: current regulation, commutation | 180 |
|     7.7.1 Torque/speed characteristic | 183 |
|     7.7.2 Shaft position sensing | 188 |
| 7.8 Solid rotors | 188 |
| Problems for Chapter 7 | 190 |
| **REFERENCES AND FURTHER READING** | 192 |
| **ANSWERS TO THE PROBLEMS** | 200 |
| **INDEX** | 202 |

# Glossary of symbols

| | | |
|---|---|---|
| $a$ | no. of parallel paths in winding | |
| $A$ | area | m² |
| $A$ | electric loading | A/m |
| $A_m$ | magnet pole area | m² |
| $\alpha$ | angular acceleration | rad/s² |
| $\alpha$ | phase angle | deg or rad |
| $\alpha$ | overlap angle | deg or rad |
| $\alpha$ | pole arc/pole-pitch ratio | |
| $\beta$ | torque angle | deg or rad |
| $\beta$ | pole arc | deg or rad |
| $B$ | flux density | T |
| $B$ | magnetic loading (p. 22) | T |
| $B_g$ | airgap flux-density (radial) | T |
| $B_r$ | remanent flux-density | T |
| $B_s$ | saturation flux-density | T |
| $C$ | commutation | |
| $C_\Phi$ | flux concentration factor | |
| $\gamma$ | slot pitch | elec deg or rad |
| $\gamma$ | phase angle defined on p. 102 | deg or rad |
| $\gamma$ | phase angle defined on p. 136 | deg or rad |
| $\gamma$ | fraction defined on p. 113 | |
| $d$ | duty-cycle (of p.w.m.) | |
| $D$ | stator bore diameter | m |
| $\delta$ | phase angle defined on p. 136 | deg or rad |
| $e$ | instantaneous e.m.f. | V |
| $e_0$ | per-unit value of $E_0$ | p.u. |
| $E$ | r.m.s. e.m.f. | p.u. |
| $E_{dw}$ | e.m.f. ascribed to web flux | $V_{r.m.s.}$ |
| $E_q$ | open-circuit e.m.f. due to magnet | $V_{r.m.s.}$ |
| $E_0$ | value of $E$ or $E_q$ at corner-point or base frequency | $V_{r.m.s.}$ |
| $\epsilon$ | chording angle | deg or rad |
| $\epsilon$ | step angle | deg or rad |
| $f$ | frequency | Hz |
| $F$ | magnetomotive force (m.m.f.) | A t |
| $F$ | mechanical force | N |
| $F_m$ | m.m.f. across magnet | A t |
| $\phi$ | flux | Wb |
| $\phi$ | power factor angle | deg or rad |
| $\Phi$ | flux | Wb |
| $\Phi_g$ | airgap flux (per pole) | Wb |

# GLOSSARY OF SYMBOLS

| | | |
|---|---|---|
| $\Phi_{M1}$ | fundamental magnet flux per pole | Wb |
| $\Phi_r$ | remanent flux of magnet | Wb |
| $\Phi_y$ | flux in one link | Wb |
| $g$ | airgap length | m |
| $g'$ | effective airgap length $K_c g$ | m |
| $g''$ | effective airgap length allowing for magnet | m |
| $g''_d$ | effective airgap length in direct-axis, allowing for magnet | m |
| $g''_q$ | effective airgap length in quadrature-axis | m |
| $H$ | magnetizing force or magnetic field strength | A t/m |
| $H_c$ | coercive force | A t/m |
| $H_{ci}$ | intrinsic coercivity | A t/m |
| $i$ | current (instantaneous) | A |
| $I$ | r.m.s. current or d.c. current | $A_{r.m.s.}$ |
| $I_c$ | controller maximum current | $A_{r.m.s.}$ |
| $J$ | current density | A/m² |
| $J$ | magnetization | T |
| $J$ | polar moment of inertia | kg m² |
| $J_m$ | motor inertia | kg m² |
| $J_L$ | load inertia | kg m² |
| $k$ | peak/continuous torque ratio (p. 27) | |

| | | |
|---|---|---|
| $k$ | armature constant (p. 66) | |
| $k$ | frequency ratio defined on p. 107 | |
| $k$ | coupling coefficient | |
| $k_{w1}$ | fundamental harmonic winding factor | |
| $k_{d1}$ | fundamental distribution factor | |
| $k_{p1}$ | fundamental pitch (chording) factor | |
| $k_{s1}$ | fundamental skew factor | |
| $k_w$ | winding factor for inductance | |
| $k_{\alpha d}$ | defined on p. 129 | |
| $k_{1ad}$ | defined on p. 130 | |
| $K_c$ | Carter's coefficient (p. 60) | |
| $l_k$ | effective length of core or keeper | m |
| $l_m$ | magnet length (in dir'n of magnetization) | m |
| $L$ | inductance | H |
| $L_a$ | aligned inductance | H |
| $L_u$ | unaligned inductance | H |
| $l$ | stack length | m |
| $\lambda$ | inductance ratio | |
| $M$ | mutual inductance | H |
| $\mu$ | permeability | H/m |
| $\mu_r$ | relative permeability | |

# GLOSSARY OF SYMBOLS

| Symbol | Description | Units |
|---|---|---|
| $\mu_{rec}$ | relative recoil permeability | |
| $n$ | gear ratio | |
| $n$ | speed | rev/sec |
| $n$ | no. of phases or phaselegs (Ch. 7) | |
| $N$ | speed | r.p.m. |
| $N$ | no. of turns | |
| $N_p$ | no. of turns per pole | |
| $N_{ph}$ | no. of series turns per phase | |
| $N_r$ | no. of rotor poles | |
| $N_s$ | no. of stator poles | |
| $N_s$ | no. of series turns per phase of sine-distributed winding or equivalent sine-distributed winding | |
| $N_1$ | no. of turns in coil 1, etc. | |
| $p$ | no. of pole pairs | |
| $P$ | permeance | Wb/A t |
| $P_{m0}$ | magnet permeance | Wb/A t |
| $P_m$ | magnet permeance including rotor leakage permeance | Wb/A t |
| $P_{r1}$ | rotor leakage permeance | Wb/A t |
| $P$ | power | W |
| PC | permeance coefficient | |
| $q$ | no. of slots per pole per phase (Ch. 5) | |
| $q$ | no. of phases (Ch. 7) | |
| $\theta$ | angular coordinate; rotor position | deg or rad |
| $\theta_D$ | dwell angle; conduction angle (of main switches) | deg or rad |
| $r_0$ | rotor slot-bottom radius | m |
| $r_1$ | rotor outside radius (Ch. 7) | m |
| $r_1$ | stator bore radius (Ch. 4–6) | m |
| $r_2$ | stator slot-bottom radius | m |
| $r_3$ | stator outside radius | m |
| $R$ | resistance (of phase winding) | Ohm |
| $R$ | reluctance | A t/Wb |
| $R_g$ | airgap reluctance | A t/Wb |
| $s$ | split ratio (p. 23) | |
| $\sigma$ | airgap shear stress | kN/m² or p.s.i. |
| $\sigma$ | half the skew angle | mech deg or rad |
| $t$ | time | s |
| $T$ | temperature | deg C |
| $T$ | torque | Nm |
| $T_a$ | average torque | Nm |
| TRV | torque per rotor volume | Nm/m³ |
| $u$ | magnetic potential | A t |
| $v$ | linear velocity | m/s |
| $v$ | voltage (instantaneous) | V |
| $V$ | voltage (d.c. or r.m.s. d.c.) | V |
| $V_c$ | controller max. voltage | $V_{r.m.s.}$ |
| $V_c$ | copper volume | m³ |

# GLOSSARY OF SYMBOLS

| | | |
|---|---|---|
| $V_m$ | magnet volume | m³ |
| $V_r$ | rotor volume | m³ |
| $w$ | web width | m |
| $w_m$ | magnet width | m |
| $W$ | energy | J |
| $W$ | conversion energy per stroke | J |
| $W'$ | coenergy | J |
| $W_f$ | stored field energy | J |
| $x$ | per-unit reactance | p.u. |
| $X$ | reactance | Ohm |
| $X_s$ | synchronous reactance | Ohm |
| $X_{s0}$ | value of $X_s$ at corner-point | Ohm |
| $X_d$ | direct-axis synchronous reactance | Ohm |
| $X_q$ | quadrature-axis synchronous reactance | Ohm |
| $X_\sigma$ | armature leakage reactance | Ohm |
| $y$ | link width | m |
| $\psi$ | flux-linkage (in Ch. 7, of phase winding) | V s or Wb t |
| $\omega$ | elec. angular velocity $=2\pi f$ (in Ch. 5–6) | elec rad/s rad/s |
| $\omega$ | mech. angular velocity (Ch. 7) | mech rad/s |
| $\omega_b$ | 'base' or 'corner-point' speed | rad/s |
| $\omega_m$ | mech. angular velocity $\omega/p$ (Ch. 5, 6) | mech rad/s |

## Subscripts

| | |
|---|---|
| a | armature-reaction |
| a | aligned |
| a | average |
| a, b, c | phases a, b, c |
| d | direct-axis |
| e | electrical, electromagnetic |
| g | airgap; airgap flux |
| L | load |
| m | mechanical |
| m, M | magnet |
| mc | maximum continuous |
| mp | peak rated |
| ph | phase |
| q | quadrature |
| r | rotor |
| s | stator |
| s | saturated |
| u | unaligned |
| w | winding factor |
| 0 | corner-point or 'base' value |
| 1 | fundamental harmonic component |
| 1, 2 | pertaining to coils 1, 2 (Ch. 4) |
| 1, 2, 3 | phases |

## Superscripts

| | |
|---|---|
| u | unsaturated |
| ~ | phasor (complex quantity) |
| ^ | peak (of sinewave) |

# 1 Introduction

## 1.1 Motion control systems

Technology is so saturated with developments in microelectronics that it is easy to forget the vital interface between electrical and mechanical engineering. This interface is found wherever mechanical motion is controlled by electronics, and pervades a vast range of products. A little consideration reveals a large and important area of technology, in which motor drives are fundamental. In Japan the term 'mechatronics' is applied to this technology, usually with the connotation of small drives. In the west the term 'motion control system' is often used for small controlled drives such as position or velocity servos. In the larger industrial range the term 'drive' usually suffices.

Many electronics engineers have the impression that the technology of motors and drives is mature, even static. But there is more development activity in drives today than at any time in the past, and it is by no means confined to the control electronics. Two important reasons for the development activity and the increasing technical variety are:

(1) Increasing use of computers and electronics to control mechanical motion. The trend towards automation demands new drives with a wide variety of physical and control characteristics.

(2) New 'enabling technology' in power semiconductors and integrated circuits, leading to the development of nonclassical motors such as brushless d.c. motors and steppers in a wide variety of designs.

## 1.2 Why adjustable speed?

Three common reasons for preferring an adjustable-speed drive over a fixed-speed motor are:

(1) energy saving;
(2) velocity or position control; and
(3) amelioration of transients.

(1) Energy saving. In developed economies about one-third of all primary energy is converted into electricity, of which about two-thirds is re-converted in electric motors and drives, mostly integral-horsepower induction motors running essentially at fixed speed. If a constant-speed motor is used to drive a flow process (such as a fan or pump), the only ways to control the flow rate are by throttling or by recirculation. In both cases the motor runs at full speed

regardless of the flow requirement, and the throttling or recirculation losses are often excessive. Similar considerations apply to the control of airflow by adjustable baffles in air-moving plant.

In such applications it is often possible to reduce average energy costs by 50 per cent or more by using adjustable-speed drives, which eliminate the throttling or recirculation loss. The adjustable-speed drive itself may be less efficient than the original fixed-speed motor, but with a drive efficiency of the order of 90 per cent this makes little difference to the overall efficiency of the process. In other words, the additional energy losses in the power electronic converter are small compared to the overall savings achieved by converting to adjustable speed. The adjustable-speed drive is more expensive, so its capital cost must be offset against energy savings. Operational advantages may also help to offset the initial cost; for example, the reduction of maintenance requirements on mechanical components.

(2) Velocity or position control. Obvious examples of speed control are the electric train, portable hand tools, and domestic washing-machine drives. In buildings, elevators or lifts are interesting examples in which not only position and velocity are controlled, but also acceleration and its derivative (jerk). Countless processes in manufacturing industry require position and velocity control of varying degrees of precision. Particularly with the trend towards automation, the technical and commercial growth in drives below about 20 kW is very vigorous. Many system-level products incorporate an adjustable-speed drive as a component. A robot, for example, may contain between three and six independent drives, one for each axis of movement. Other familiar examples are found in office machinery: positioning mechanisms for paper, printheads, magnetic tape, and read/write heads in floppy and hard disk drives.

FIG. 1.1. (a) Food processor motor with the associated switched reluctance rotor (centre) having 2.5–3 times the torque of the universal motor armature (left) over the whole speed range. Courtesy Switched Reluctance Drives Ltd., Leeds.

# WHY ADJUSTABLE SPEED?

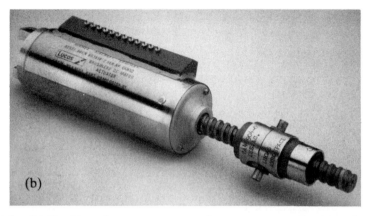

FIG. 1.1. (b) Brushless d.c. actuator motor for aerospace applications. Courtesy Lucas Engineering and Systems Ltd., Solihull.

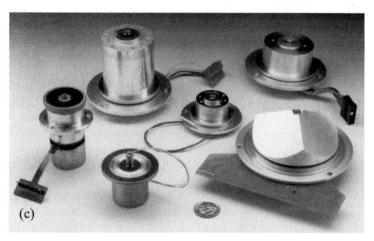

FIG. 1.1. (c) Brushless computer disk drive motors. Courtesy Synektron Corporation, Portland, Oregon, USA.

(3) Amelioration of transients. The electrical and mechanical stresses caused by direct-on-line motor starts can be eliminated by adjustable-speed drives with controlled acceleration. A full adjustable-speed drive is used in this situation only with very large motors or where the start–stop cycles are so frequent that the motor is effectively operating as a variable speed drive. Most soft-starting applications are less onerous than this, and usually it is sufficient (with a.c. motors) to employ series SCRs (or triacs with smaller motors) which 'throttle' the starting current to a controlled value, and are bypassed by a mechanical switch when the motor reaches full speed. Series control of induction motors is inefficient, produces excessive line harmonics, and is not

4   INTRODUCTION

FIG. 1.1. (d) Industrial general-purpose adjustable-speed switched reluctance drive. Courtesy Tasc Drives Ltd., Lowestoft.

very stable. The soft-starter is less expensive than a full adjustable-speed drive, which helps to make it economical for short-time duty during starting; but it is not ideal for continuous speed control.

*1.2.1 Large versus small drives*

There are marked design differences between large and small drives. Large motors are almost always chosen from one of the classical types: d.c. commutator (with wound field); a.c. induction; or synchronous. The main reasons are the need for high efficiency and efficient utilization of materials and the need for smooth, ripple-free torque. In small drives there is greater variety because of the need for a wider range of control characteristics. Efficiency and materials utilization are still important, but so are control characteristics such as torque/inertia ratio, dynamic braking, and speed range.

There are also breakpoints in the technology of power semiconductors. At the highest power levels (up to 10 MW) SCRs (thyristors) and GTOs (gate turn-off thyristors) are the only devices with sufficient voltage and current capability. Naturally-commutated or load-commutated converters are preferred, because of the saving in commutation components and for reliability and efficiency reasons. In the medium power range (up to a few hundred kW) forced commutation is a design option and bipolar transistors, Darlington transistors, and GTOs are popular. At low powers (below a few kW) the power MOSFET is attractive because it is easy to switch at high chopping

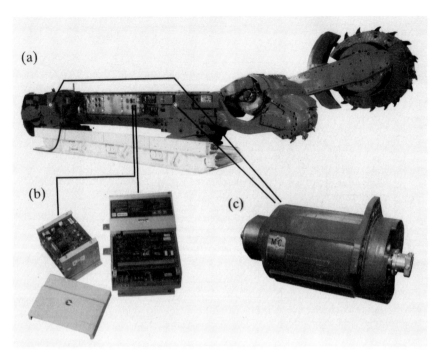

FIG. 1.2. Thyristor controlled d.c. drive applied to coal shearer. (a) The drive rated at 75 kW, provides the traction to drive the Shearer along the coal face at a speed controlled automatically to maintain the main a.c. cutter motors at optimum load appropriate to the hardness of the coal. (b) The main drive module incorporates a heavy metal base plate which acts as a heat sink for the devices and bolts directly to a water cooled wall within the explosion proof enclosure housed within the machine structure. (c) The d.c. motors (two) have their armatures connected in series to ensure load sharing between two traction units mounted at the two ends of the machine. The motors themselves are specially designed, fully compensated explosion proof machines with water cooled armatures to ensure adequate cooling in minimum space. Courtesy Anderson Strathclyde PLC, Glasgow; Control Techniques PLC, Newtown, Powys; and David McClure Limited, Stockport.

frequencies and ideally suited to the needs of various pulse-width-modulated (p.w.m.) converters. New devices such as the insulated-gate transistor (IGT) are also making progress at low power levels (Baliga 1987), and more recently in the medium power range as well.

## 1.3 Structure of drive systems

The general structure of a 'motion control system' or 'drive' is shown in Fig. 1.3. The system is integrated from four distinct elements:

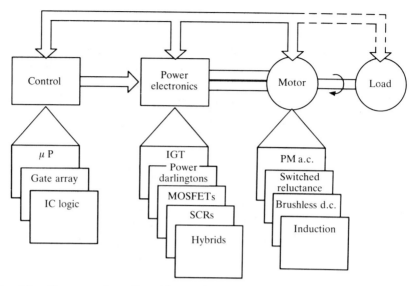

FIG. 1.3. Structure of an adjustable-speed drive system. Few motion-control systems can be made cost-effective without careful consideration of alternative component technologies and their integration into a complete system.

(1) the load;
(2) the motor;
(3) the power electronic converter; and
(4) the control.

The range of modern motion-control applications is virtually unlimited. Any random list of twenty provides a rich technical inventory: aerospace actuators; washing machines; computer disk and tape drives; printer platens and printheads; inertial guidance systems; adjustable-speed pumps, blowers, and fans; locomotive and subway traction; automatic machine tools and machining centers; servodrives and spindle drives; robots; automotive auxiliaries; refrigeration and air-conditioning drives; and many others.

Loads have widely differing requirements. The commonest requirement is for speed control, with varying degrees of precision and accuracy. Position control is of increasing importance, particularly in automated plants and processes, and in office machinery and computer peripherals. In some cases it is the steady-state operation that is most important, for example in air-conditioning and pump drives. In other cases, such as in robots, tape drives, and actuators, dynamic performance is important because of the need to minimize the time taken to perform operations or effect control changes. In these cases the torque/inertia ratio of the motor is an important parameter. In automotive applications the prime requirements are low cost and low noise.

Efficiency is important in motors that run continuously, e.g. heater blowers, but not in intermittent-duty motors such as window-winders.

Table 1 provides a checklist of some of the performance parameters and constraints to consider in the application of an adjustable-speed drive.

**Table 1.1.** General application check-list for adjustable speed drives

1. Compliance with national, EC, USA, and industry standards
2. Maximum continuous power or torque requirement
3. Forward/reverse operation
4. Motoring and/or braking operation
5. Dynamic or regenerative braking
6. Overload rating and duration
7. Supply voltage (a.c. or d.c.) and frequency (a.c.)
8. Type of control: speed, position, etc.
9. Precision required in controlled speed or position
10. Programmability: speed and/or position profiles, start/stop ramps, etc.
11. Interfaces with control and communications equipment, plant computers, etc.
12. Dynamic requirements: torque/inertia ratio, acceleration and deceleration capability
13. Gearbox or direct drive; optimal gear ratio
14. Reliability and redundancy of components
15. Protection arrangements for mechanical and electronic failures and abnormal conditions
16. Maximum level of acoustic noise
17. Maximum level of electromagnetic radiation (EMI) (conducted and radiated)
18. Maximum levels of harmonics in the supply and the motor
19. Maintenance; spare parts; provision for expansion or rearrangement of plant
20. Environmental factors: indoor/outdoor installation; enclosure; temperature; humidity; pollution; wind and seismic factors; type of coolant

## 1.4 New technology

Several new technologies are contributing to the development of motion control systems.

### 1.4.1 Digital electronics

It would be hard to overstate the importance of microelectronics in motion control. At the 'heavy' end of the spectrum are the multiple drives found in steel rolling mills, paper mills, and other heavy process plants, where it is normal to coordinate the motion of all the shafts by means of a computer or a network of computers, some of which may be quite large. At the light end of the spectrum are the drives found in office machinery and small computers, where custom integrated circuits and gate arrays are common. Between these extremes there are many microprocessor-controlled systems of all levels of complexity.

The first functions implemented with microprocessors were low-speed functions such as monitoring and diagnostics, but digital control has penetrated from outer position loops through the intermediate velocity loop and even into high-speed current regulators. The development of 'field-oriented' or 'vector' control for a.c. induction and synchronous motor drives would not have been practical without the microprocessor. This technique, which is based on the reference-frame transformations that date back to Park and others, permits the outer control loops of a.c. and d.c. drives to be the same, both in hardware and software, and improves the dynamic performance of the a.c. drive.

Because of the limitations and costs of current-regulation schemes and their sensors, it is more usual in a.c. drives to find p.w.m. control applied to the voltage rather than to the current, and innumerable algorithms have been developed to do this. Some of these use custom ICs or gate arrays. A few microprocessors are fast enough to perform real-time p.w.m. control. In the future, the concentration of control power in microelectronic devices may facilitate the direct regulation of instantaneous current and torque, instead of the average torque normally calculated in the synchronously rotating reference frame; but this cannot be accomplished without fast power switches and excess volt-ampere capacity.

Digital control is standard in axis controllers for CNC machine tools and related motion-control applications, invariably with communications interfaces such as RS232, IEEE488, MULTIBUS and so on. Such controllers may be used with d.c., a.c., or brushless drives.

### 1.4.2 Power integrated circuits

Conventional ICs are limited to standard TTL or CMOS voltage levels and require interface and level-shifting circuitry between them and the power semiconductor switches they control. Hybrids are used in this application, but more recently there have appeared several 'power ICs' (PICs) with voltage ratings of 40–100 V and one or two with much higher voltage ratings ranging up to 500 V (Steigerwald *et al.* 1987). Several of these chips can source and sink currents of 2 A at 40–50 V, for example Sprague's UDN2936W and Unitrode's UC3620, providing full phaselegs or bridges with many additional control and protective functions. National's LM621 brushless motor driver provides 35 mA outputs at 40 V in a full three-phase bridge configuration and includes not only the decoding (commutation logic) for a Hall-effect shaft sensor but also a PWM current-regulating facility, as well as 'lock-out' protection (to prevent shoot-through or shorting the d.c. supply) and undervoltage and overcurrent protection. Motorola's MC33034 is similar, with 50 mA outputs and an integral window detector with facilities for speed feedback. General Electric's High-Voltage Integrated Circuit provides source and sink driver signals at 500 V d.c. for a single 'totem-pole' phaseleg. Other chips use dielectric isolation to achieve comparable voltage ratings.

The power IC permits a massive reduction in component part count and can short-circuit much of the design task. It brings savings in converter weight and volume, along with improvements in thermal management, EMI, and reliability.

Another circuit function that has been 'integrated' recently is current sensing. In Motorola's SENSEFET and International Rectifier's HEXSense power MOSFET, a current-mirror is formed from a small number of cells and provides a current-sensing signal to a user-supplied resistor. Voltage ratings up to 500 V are available at 10 A.

### 1.4.3 Power semiconductor devices

While GTO thyristors have been widely adopted for large drives (and in Europe even down to 1 kW), the steady progress in FET and bipolar junction transistors (BJTs) has made them the natural choice in many applications below 100 kW. PWM switching frequencies above the audible range are common, and converter efficiency and reliability are very high.

Available from Japan and the USA, the IGBT now has voltage ratings up to 1 kV with current capability of up to 200 A, and combines the power-handling characteristics of a BJT with the controllability of a FET.

### 1.4.4 New magnetic materials

The permanent-magnet industry has had sustained success in developing improved magnet characteristics. The most recent major advance is neodymium–iron–boron, pioneered particularly by Sumitomo and General Motors. At room temperature NdFeB has the highest energy product of all commercially available magnets. Ceramic (ferrite) magnets have also been steadily improved (see Chapter 3).

### 1.4.5 CAD and numerical analysis in design

Motor design has been computerized since the early days of computers, initially with the coding of well established design procedures. More recently electromagnetic field analysis has emerged from the academic world into the design office as a tool for optimizing designs, particularly with respect to the efficient utilization of materials and optimization of geometry. The most popular methods for analysis are based on the finite-element method, and there are few design offices not equipped with it in one form or another. Several commercial packages exist for magnetostatic nonlinear problems in two dimensions (usually in a transverse cross-section of the motor). Three-dimensional packages are under intense development and several programs are available to handle eddy-current problems in two dimensions.

The finite-element technique is an analysis tool rather than a design tool, and it suffers from certain limitations when applied to motor design. It requires detailed input data and the results need skilled interpretation. It is accurate only in idealized situations where parasitic effects have been removed. It is too slow to be cost-effective as part of a synthesizing CAD package, and is likely to remain so for some time. It is most useful in helping to understand a theoretical problem that is too difficult for conventional analysis, and in this role it has undoubtedly helped to refine many existing motor designs and improve some new ones.

The primary problems in motor design are not simply electromagnetic but require a synthesizing approach to materials utilization and design-for-manufacture. This multidisciplinary problem includes not only electromagnetic analysis but heat transfer and mechanical design as well. The situation is more complicated in adjustable-speed drives where the supply waveforms are 'switchmode' rather than pure sinewaves or d.c. In these cases simulation may be necessary to determine the expected performance of a given design over a wide range of operating conditions.

In view of this it is surprising that synthesizing CAD software for motor design is rather rare, even at the elementary or 'sizing' level. This is likely to change rapidly as a result of the development and widespread adoption of PCs and workstations. In the longer term, there are possibilities for a revival in optimization techniques and the application of artificial intelligence.

For system simulation there are several software packages, such as SIMNON from the University of Lund, Sweden; Control-C; ACSL; the Electromagnetic Transients Program (EMTP); and others. Suitably modified and extended, some of these packages permit the simulation of quite detailed motor models as well as their controls. They may be used for the development of control algorithms that are subsequently programmed in a microprocessor or gate array.

### 1.4.6 *Other contributing technologies*

Plastics and composite materials find many applications in motors. Fans, slot liners and wedges, endbells and covers, and winding supports are the commonest, but moulded slot insulation and encapsulation of rotors are newer possibilities. In brushless motors designed for high peripheral speeds, Kevlar or glass banding can be used to retain surface-mounted magnets.

Motor drives often require transducers for control and protection, and there has been progress in current-sensor and shaft position sensor technology. In particular the linearity and temperature-independence of Hall-effect current sensors has improved greatly, and it is common to mount these devices in the same package, or on the same printed-circuit card, as the driver stage of the power electronics in small drives. For larger drives flux-nulling current sensors can be used with bandwidths of up to several kHz and isolation at least as good as that of a CT.

In brushless drives the commutation signals are often derived from three Hall sensors, activated either by the rotor magnet or by a separate magnet ring. Alternatively, optical interrupters may be used with a shaft-mounted slotted disk. At high speeds the commutation sensor can be used to generate a speed signal via a frequency-to-voltage conversion. For motion control systems and servo-quality drives separate velocity and position transducers usually have to be used. For such systems the resolver is attractive because of its ruggedness, resolution, and its ability to provide accurate absolute position and velocity signals from one sensor. 'R-to-D' (resolver-to-digital) converters are available with tracking capability up to more than 40 000 r.p.m.

## 1.5 Which motor?

The proliferation of new ideas, materials, and components obviously generates many opportunities but also complicates the question, what is the best drive for a particular job? We can perhaps address this by attempting to trace the evolution of the different motor types in such a way as to bring out their most important advantages and disadvantages. It is the motor that determines the characteristics of the drive, and it also determines the requirements on the power semiconductors, the converter circuit, and the control.

### 1.5.1 Evolution of motors

The evolution of brushless motors is shown in Fig. 1.4. Row 1 contains the three 'classical' motors: d.c. commutator (with wound field); a.c. synchronous; and a.c. induction. The term 'classical' emphasizes the fact that these motors satisfy three important criteria:

(1) they all produce essentially constant instantaneous torque (i.e., very little torque ripple);
(2) they operate from pure d.c., or a.c. sinewave supplies, from which
(3) they can start and run without electronic controllers.

The classical motors of Row 1 are readily coupled to electronic controllers to provide adjustable speed; indeed it is with them that most of the technical and commercial development of power electronic control has taken place. Together with the PM d.c. commutator motor in Row 2 and the series-wound a.c. commutator motor or 'universal' motor, the Row 1 motors account for the lion's share of all motor markets, both fixed-speed and adjustable-speed, even

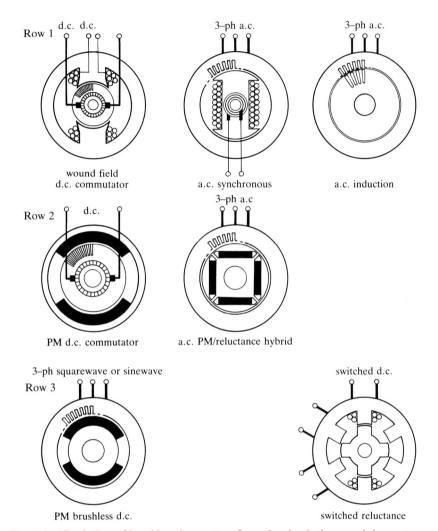

FIG. 1.4. Evolution of brushless d.c. motors from the classical a.c. and d.c. motors.

though they represent only a minority of the many different principles of electromechanical energy conversion on which motor designs may be based. By contrast, the nonclassical motors are essentially confined to specialist markets and until recently, few of them have been manufactured in large numbers. Table 1.2 is a classification of some common types of motor according to these criteria.

The motors in Row 2 are derived from those in Row 1 by replacing field windings with permanent magnets. The synchronous motor immediately becomes brushless, but the d.c. motor must go through an additional transformation, from Row 2 to Row 3 with the inversion of the stator and rotor, before the brushless version is achieved. The induction motor in Row 1 is, of course, already brushless in its 'cage' version, but not in its wound-rotor or slip-ring version. The brushless motors are those on the diagonal of Fig. 1.4, together with the switched reluctance motor, which cannot be derived from any of the other motors. Its placement in Fig. 1.4 reflects the fact that it has properties in common with all the motors on the diagonal, as will be seen later. Obviously the emphasis in this book is on the brushless motors, but excluding detailed treatment of the induction motor, which is treated in detail in other texts (e.g. Leonhard 1985, Bose 1986, 1987). Stepper motors are also excluded (see Kuo 1979, Kenjo 1985, Acarnley 1982).

### 1.5.2 The d.c. commutator motor

The traditional d.c. commutator motor with electronically adjustable voltage has always been prominent in motion control. It is easy to control, stable, and requires relatively few semiconductor devices. Developments in electronics have helped to keep it competitive in spite of efforts to displace it with a.c. drives.

Many objections to the commutator motor arise from operational problems associated with the brushgear. It is not that brushgear is unreliable—on the contrary, it is reliable, well-proven, and 'forgiving' of abuse—but commutator speed is a limitation, and noise, wear, RFI, and environmental compatibility can be troublesome. The space required for the commutator and brushgear is considerable, and the cooling of the rotor, which carries the torque-producing winding, is not always easy.

### 1.5.3 The PM d.c. commutator motor

In small d.c. commutator motors, replacing the field winding and pole structure with permanent magnets usually permits a considerable reduction in stator diameter, because of the efficient use of radial space by the magnet and the elimination of field losses. Armature reaction is usually reduced and commutation is improved, owing to the low permeability of the magnet. The loss of field control is not as important as it would be in a larger drive, because it can be overcome by the controller and in small drives the need for field weakening is less common anyway. The PM d.c. motor is usually fed from an adjustable voltage supply, either linear or pulse-width modulated.

In automotive applications the PM d.c. motor is well entrenched because of its low cost and because of the low-voltage d.c. supply. Of course it is usually applied as a fixed-speed motor or with series-resistance control. Even here, however, there is a potential challenge from brushless motor drives in the future, arising from the combination of very high reliability requirements and the development of 'multiplex' wiring systems.

**Table 1.2.** Motors

| Motor | Type of supply | Rotor position feedback | Typical application |
|---|---|---|---|
| **d.c. commutator motors** | | | |
| (a) Wound field | Controlled or fixed d.c. | N | Integral-hp industrial drives and traction; Steel, paper machinery Automotive and aircraft auxiliaries; Small servo and speed-control systems in a wide variety of forms |
| (b) Permanent-magnet | Controlled or fixed d.c. | N | |
| d.c. Homopolar motors | Controlled or fixed d.c. | N | Ship propulsion specials |
| d.c. Brushless motors | 120° square waves of alternating polarity or three-phase sinewave a.c. | Y | Motion-control systems; servodrives Aerospace actuators Computer peripherals; office machinery Small fan and pump drives |
| (a) Internal rotor | | | |
| (b) Internal stator | | | |
| Universal or a.c. commutator motors | One-phase a.c.; can be controlled by series SCR or triac | N | Domestic appliances; portable tools |
| **Induction motors** | | | |
| (a) Cage type, three-phase | Three-phase sinewave a.c. or six-step or p.w.m. a.c. | N | Pumps, fans, blowers, compressors and general industrial speed-controlled drives |
| (b) Cage type, single-phase | One-phase sinewave a.c.; can be controlled by series SCRs or triacs | N | Low-cost, low-power industrial and domestic appliance drives |
| (c) Wound rotor, three-phase | Three-phase sinewave a.c. | N | High-power industrial drives with limited speed range and/or high starting torque |

# WHICH MOTOR?

| | | | |
|---|---|---|---|
| **Synchronous motors** | | | |
| (a) Wound field/slip ring | Three-phase sinewave a.c. or six-step voltage or current-source inverter | N | Very large compressor and fan drives |
| (b) Brushless exciter | | Y | |
| (c) Permanent-magnet | Three-phase sinewave a.c. or p.w.m. a.c. | Y | Low integral-hp industrial drives; fibre spinning |
| **Reluctance motors** | | | |
| (a) Synchronous reluctance (line-start) | Three-phase sinewave a.c. or six-step or p.w.m. a.c. | N | Inverter-fed spinning machinery and other multiple synchronous drives |
| (b) Synchronous reluctance (cageless) | Three-phase p.w.m. a.c. | Y | |
| (c) Switched reluctance | Switched d.c. | Y | Low-cost brushless drive applications with wide speed range; domestic appliances, industrial drives up to 50–100 kW; aerospace applications |
| (d) One-phase reluctance | Various; usually switched d.c. | N | Very small synchronous drives, actuators, watches |
| **Stepper motors** | | | |
| (a) VR, single-stack | Switched d.c. | N | Printers; plotters; position control |
| (b) VR, multiple-stack | | | |
| (c) Permanent-magnet | | | |
| (d) Hybrid | | | |
| Hysteresis motors | Three-phase or one-phase sinewave a.c. or p.w.m. a.c. | N | Turntables |
| Inductor motors | Three-phase sinewave a.c. or p.w.m. a.c. | N | High-speed applications |

## 1.5.4 The a.c. induction motor drive

In very large drives a.c. induction or synchronous motors are preferred because of the limitations of commutation and rotor speed in d.c. motors.

Slip is essential for torque production in the induction motor, and it is impossible, even in theory, to achieve zero rotor losses. This is one of the chief limitations of the induction motor, since rotor losses are more difficult to remove than stator losses.

The efficiency and power factor of induction motors falls off in small sizes because of the natural laws of scaling, particularly at part load. As a motor of given geometry is scaled down, if all dimensions are scaled at the same rate the m.m.f. required to produce a given flux-density decreases in proportion to the linear dimension. But the cross-section available for conductors decreases with the square of the linear dimension, as does the area available for heat transfer. This continues down to the size at which the mechanical airgap reaches a lower limit determined by manufacturing tolerances. Further scaling-down results in approximately constant m.m.f. requirements while the areas continue to decrease with the square of the linear dimension. There is thus an 'excitation penalty' or 'magnetization penalty' which becomes rapidly more severe as the scale is reduced. It is for this reason that permanent magnets are so necessary in small motors. By providing flux without copper losses, they directly alleviate the excitation penalty.

The induction motor is indeed 'brushless' and can operate with simple controls not requiring a shaft position transducer. The simplest type of inverter is the six-step inverter. With no shaft position feedback, the motor remains stable only as long as the load torque does not exceed the breakdown torque, and this must be maintained at an adequate level by adjusting the voltage in proportion to the frequency. At low speeds it is possible for oscillatory instabilities to develop. To overcome these limitations a range of improvements have been developed including slip control and, ultimately, full 'field-oriented' or 'vector' control in which the phase and magnitude of the stator currents are regulated so as to maintain the optimum angle between stator m.m.f. and rotor flux. Field orientation, however, requires either a shaft position encoder or an in-built control model whose parameters are specific to the motor, and which must be compensated for changes that take place with changing load and temperature. Such controls are complex and expensive, and cannot be justified in very small drives, even though excellent results have been achieved in larger sizes (above a few kW).

In the fractional and low integral-horsepower range the complexity of the a.c. drive is a drawback, especially when dynamic performance, high efficiency, and a wide speed range are among the design requirements. These requirements cannot be met adequately with series- or triac-controlled induction motors, which are therefore restricted to applications where low cost is the only criterion.

Together these factors favour the use of brushless PM drives in the low power range.

### 1.5.5 The brushless d.c. PM motor

The smaller the motor, the more sense it makes to use permanent magnets for excitation. There is no single 'breakpoint' below which PM brushless motors outperform induction motors, but it is in the 1–10 kW range. Above this size the induction motor improves rapidly, while the cost of magnets works against the PM motor. Below it, the PM motor has better efficiency, torque per ampere, and effective power factor. Moreover, the power winding is on the stator where its heat can be removed more easily, while the rotor losses are extremely small. These factors combine to keep the torque/inertia ratio high in small motors.

The brushless d.c. motor is also easier to control, especially in its 'squarewave' configuration (see Chapter 4). Although the inverter is similar to that required for induction motors, usually with six transistors for a three-phase system, the control algorithms are simpler and readily implemented in 'smartpower' or PICs.

### 1.5.6 The brushless PM a.c. synchronous motor

In Row 2 of Fig. 1.4 the brushless synchronous machine has permanent magnets instead of a field winding. Field control is again sacrificed for the elimination of brushes, sliprings, and field copper losses. This motor is a 'classical' salient-pole synchronous a.c. motor with approximately sine-distributed windings, and it can therefore run from a sinewave supply without electronic commutation. If a cage winding is included, it can self-start 'across-the-line'.

The magnets can be mounted on the rotor surface (see Chapter 5) or they can be internal to the rotor. The interior construction simplifies the assembly and relieves the problem of retaining the magnets against centrifugal force. It also permits the use of rectangular instead of arc-shaped magnets, and usually there is an appreciable reluctance torque which leads to a wide speed range at constant power.

The PM synchronous motor operates as a synchronous reluctance motor if the magnets are left out or demagnetized. This provides a measure of fault-tolerance in the event of partial or total demagnetization through abnormal operating conditions. It may indeed be built as a magnet-free reluctance motor, with or without a cage winding for starting 'across-the-line'. However, the power factor and efficiency are not as good as in the PM motor (see Chapter 6).

In larger sizes the brushless synchronous machine is sometimes built with a brushless exciter on the same shaft, and a rotating rectifier between the exciter and a d.c. field winding on the main rotor. This motor has full field control. It is capable of a high specific torque and high speeds.

All the motors on the diagonal of Fig. 1.4 share the same power circuit topology (three 'totem-pole' phaselegs with the motor windings connected in star or delta to the midpoints). This gives rise to the concept of a family of motor drives providing a choice of motors and motor characteristics, but with a high degree of commonality in the control and power electronics and all the associated transducers. The trend towards integrated phaselegs, or indeed complete three-phase bridges, with in-built control and protection circuitry makes this concept more attractive. This family of drives covers a wide range of requirements, the main types being the conventional brushless d.c. (efficient in small sizes with good dynamics); the interior-magnet synchronous motor (wide speed range); the synchronous reluctance motor (free from magnets and capable of very high speeds or high-temperature operation); and the induction motor. It should be noted that all these drives are essentially 'smooth-torque' concepts with low torque ripple.

A major class of motors not included in Fig. 1.4 is the stepper motor. Steppers are always brushless and almost always operate without shaft position sensing. Although they have many properties in common with synchronous and brushless d.c. motors they cannot naturally be evolved from the motors in Fig. 1.4. By definition they are pulsed-torque machines incapable of achieving ripple-free torque by normal means. Variable-reluctance (VR) and hybrid steppers can achieve an internal torque multiplication through the use of multiple teeth per stator pole and through the 'vernier' effect of having different numbers of rotor and stator teeth. Both these effects work by increasing the number of torque impulses per revolution, and the price paid is an increase in commutation frequency and iron losses. Steppers therefore have high torque-to-weight and high torque-to-inertia ratios but are limited in top speed and power-to-weight ratio. The fine tooth structure requires a small airgap, which adds to the manufacturing cost. Beyond a certain number of teeth per pole the torque gain is 'washed out' by scale effects that diminish the variation of inductance on which the torque depends. Because of the high magnetic frequency and the effect of m.m.f. drop in the iron, such motors require expensive lamination steels to get the best out of them.

The switched reluctance motor or variable-reluctance motor is a direct derivative of the single-stack VR stepper, in which the current pulses are phased relative to the rotor position to optimize operation in the 'slewing' (continuous rotation) mode. This usually requires a shaft position transducer similar to that which is required for the brushless d.c. motor, and indeed the resulting drive is like a brushless d.c. drive without magnets. With this form of control the switched reluctance motor is not a stepper motor because it can produce continuous torque at any rotor position and any speed. There is still an inherent torque ripple, however. The switched reluctance motor suffers the same 'excitation penalty' as the induction motor and cannot equal the efficiency or power density of the PM motor in small framesizes.

When the classical motors are interfaced to switchmode converters (such as rectifiers, choppers, and inverters) they continue to respond to the average voltage (in the case of d.c. motors) or the fundamental voltage (in the case of a.c. motors). The harmonics associated with the switching operation of the converter cause parasitic losses, torque ripple, and other undesirable effects in the motors, so that de-rating may be necessary. The nonclassical motors are completely dependent on the switchmode operation of power electronic converters. In steppers it is acceptable for the torque to be pulsed, but in most brushless drives the challenge is to design for smooth torque even though the power is switched.

# 2 Principles of sizing, gearing, and torque production

## 2.1 Sizing an electric motor

If we think of the torque as being produced at the rotor surface, Fig. 2.1, we immediately get the idea of 'torque per unit rotor volume' (TRV):

$$\text{TRV} = \frac{T}{V_r} = \frac{T}{\pi r^2 l}$$

It is very common to be given the torque requirement of a load, and asked to design a motor to meet it. An initial idea of the TRV can be used to estimate the rotor volume as the starting-point of a detailed design. The TRV is well suited for this purpose, and it is perhaps useful to examine why this is so before quoting some typical values for particular motor types.

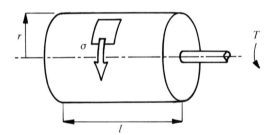

FIG. 2.1. Torque per unit rotor volume (TRV) and airgap shear stress.

### 2.1.1 Airgap shear stress

The torque can be related to an average shear stress at the rotor surface. Consider one square unit of area on the rotor surface, Fig. 2.1. If the average shear stress is $\sigma$ then the torque is given by

$$T = 2\pi r^2 l \sigma.$$

From this and the preceding equation,

$$\text{TRV} = 2\sigma.$$

The units of TRV are kN m/m³ and of $\sigma$, kN/m². In imperial units, $\sigma$ is usually expressed in p.s.i., and TRV in in lb/in³. A value of 1 p.s.i. (6.9 kN/m²) corresponds to a TRV of 2 in lb/in³ (13.8 kN m/m³).

# SIZING AN ELECTRIC MOTOR

In a consistent set of units the TRV is twice the shear stress. The shear stress is related to the levels of magnetic and electric excitation or the magnetic and electric 'loadings'. Consider a rectangular area in the airgap near the rotor surface, Fig. 2.2, through which the radial flux-density is $B$. Assume that the flux is established by an independent excitation source (such as a permanent magnet) fixed to the rotor. Let the current flowing across the small rectangular area be $I$ amperes, and assume that this is fixed to the stator. The current flows in the axial direction. The electric loading $A$ can be defined to be equal to $I/y$. The tangential force produced by the interaction of the flux and the current is $F = BIz = BAyz$, giving $\sigma = BA$. That is, the shear stress is equal to the product of the magnetic and electric loadings. Conventional definitions of magnetic and electric loadings used with particular machines usually give rise to a constant of proportionality in this relationship, i.e. $\sigma = kBA$, but it does not affect the fundamental significance of the shear stress.

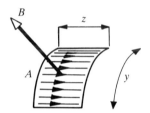

FIG. 2.2. Relationship between airgap shear stress and electric and magnetic loadings.

Consider the example of an induction motor, in which $B$ and $A$ are both sine-distributed:

$$A(\theta) = \hat{A} \sin \theta$$

$$B(\theta) = \hat{B} \sin(\theta - \alpha).$$

Note that $A$ and $B$ are not necessarily orthogonal: their axes are separated in space by the angle $\alpha$. In a d.c. motor they are maintained orthogonal by the commutator, and in a brushless d.c. motor the orthogonality is maintained, in an average sense, by electronic commutation. In the induction motor orthogonality can be forced by control means such as 'field orientation', but then $A$ must be regarded as due to the 'torque' component of stator current, not the magnetizing component. In an ordinary induction motor $\alpha$ is less than 90°.

If the electric loading is defined as

$$A = \frac{3 N_{ph} I}{2 \pi r}$$

it can be shown that

$$\hat{A} = 2\sqrt{2} k_{w1} A$$

where $k_{w1}$ is the fundamental winding factor (see Chapter 5). Also

$$\hat{B} = \frac{\pi}{2} B.$$

The product integrated over the rotor surface gives

$$\mathrm{TRV} = \frac{1}{2} \hat{A} \hat{B} \cos \alpha = \frac{\pi}{\sqrt{2}} k_{w1} A B \cos \alpha.$$

As an example, in a 20 kW induction motor $A$ might have a value of 20 A/mm (20 kA/m), and $B$ of 0.5 T; with a winding factor of 0.94 and $\cos \alpha = 0.8$, this gives TRV = 16.7 kN m/m³ and $\sigma = 8.35$ kN/m² or 1.2 p.s.i.

It is of interest to relate the electric loading to the current density in the slots. With a slot depth of 15 mm, a slot fill factor of 40 per cent, and a tooth width/pitch ratio of 0.5, the current density is

$$J = \frac{A}{0.4 \times 0.5 \times 15} = 6.6 \text{ A/mm}^2.$$

The fundamental interaction between flux and current in the production of torque is implicit in the formula $\sigma = kBA$. Note that $B$ and $A$ are both densities, respectively of flux and current. It is useful to understand how these densities are defined and how they affect the size and shape of the machine.

In conventional motors the flux crosses the airgap radially and it is natural to evaluate the magnetic loading as the average radial flux-density in the airgap. Its value is limited by the available m.m.f. of the excitation source. The m.m.f. required depends very largely on the length of the airgap and the saturation characteristics of the steel on both the rotor and the stator. In most motors, if the flux density in the stator teeth exceeds about 1.8 T then the excitation m.m.f. requirement becomes too great to be provided economically, because of the sharp increase in magnetic potential gradient ($H$) in the steel. In slotted structures, the flux density in the teeth is typically twice that in the airgap. Therefore, $\hat{B}$ is normally limited in most motors to a value around 0.9 T (peak), which gives $B = 0.57$ T. Saturation of the rotor or stator yoke ('back-iron') is a further potential cause of excessive excitation m.m.f. requirement.

The electric loading is a question of how many amperes can be packed together in a circumferential width $y$. Basically this is limited by the slot fill factor, the depth of slot, and the current density.

It is interesting at this point to see why it is the rotor volume and not its surface area that primarily determines the torque capability or 'specific output'. As the diameter is increased, both the current and the flux increase if the electric and magnetic loadings are kept the same. Hence the diameter (or radius) appears squared in any expression for specific output. On the other hand, if the length is increased, only the flux increases, not the current. Therefore the length appears linearly in the specific output. Thus the specific output is proportional to $D^2 l$ or rotor volume.

In practice as the diameter is increased, the electric loading can be increased also, because the cooling can be made more efficient without reducing the efficiency. Consequently the specific output (TRV) increases faster than the rotor volume.

Although it is theoretically possible to write one general equation from which the torque of any electric motor can be calculated, in practice a different torque equation is used for every different type of motor. Only in certain cases is it possible to discern in this equation an explicit product of flux and current, or even of quantities directly related to them. For example, in the d.c. commutator motor the torque is given by

$$T = k \phi I_a.$$

Here the current-flux product is very obvious. In the case of rotating-field a.c. machines the classical torque equations do not contain this product explicitly. However, the recent development of 'field-oriented' or 'vector' controls has necessitated the transformation of the classical equations into forms wherein the flux and current may appear explicitly in a scalar or vector product. Basically this transformation is possible because the magnetic field and the stator current waves rotate in synchronism. In the d.c. commutator machine and in a.c. rotating-field machines, the relationship between $\sigma$ and the magnetic and electric loadings can be meaningfully interpreted.

In the case of doubly-salient motors such as the switched reluctance motor and stepper motors, the transformation is not possible, and the torque cannot be expressed as the explicit product of a flux and a current. In this case, while it is still possible to define and use $\sigma$ and $A$, the magnetic loading cannot be meaningfully defined. However, the TRV can still be used for initial sizing.

## 2.1.2 Torque per unit stator volume

So far we have restricted attention to the torque per unit rotor volume, on the grounds that the TRV fixes the basic size of the motor. The stator volume results from the detailed calculations that follow from the sizing of the rotor. For very rough estimation of overall size, a typical value of 'split ratio' (i.e., rotor/stator diameter ratio) can be used. If $s$ is the split ratio, then

$$\text{Stator volume} = \frac{\text{Rotor volume}}{s^2}.$$

A typical value of split ratio for an a.c. motor is in the range 0.55–0.65. For steppers and switched reluctance motors rather smaller values are found. For PM d.c. commutator motors the value is usually somewhat higher.

The best way to acquire typical practical values of $\sigma$ or TRV is by experience. An engineer who is familiar with a particular design of motor will have built and tested several, and the test data provides values of TRV correlated with temperature rise, electric and magnetic loadings, etc. Table 2.1 provides some guidance as to the values that might be expected in practice. The values quoted here relate to the continuous rating. Peak ratings may exceed these values by 2–3 times, depending on the duration and other factors.

**Table 2.1.** Typical TRV and $\sigma$ values for common motor types

| Motor type and size | $\sigma$ (p.s.i.) | TRV (kN m/m$^3$) |
|---|---|---|
| Fractional TEFC industrial motors | 0.1–0.3 | 1.4–4 |
| Integral TEFC industrial motors | 0.5–2 | 15–30 |
| High-performance industrial servos | 1.5–3 | 20–45 |
| Aerospace machines | 3–5 | 45–75 |
| Very large liquid-cooled machines (e.g. turbine generators) | 10–15 | 130–220 |

The TRV determines the volume of the rotor but not its shape. To estimate the rotor diameter and length separately, a length/diameter ratio should be specified. A value around 1 is common; however, it is also common to design motors of different ratings using the same laminations but with different stack lengths. The length/diameter ratio may then vary over a range of 3:1 or more. Very large length/diameter ratios are undesirable because of inadequate lateral stiffness, but may be used where a very high torque/inertia ratio is desired, or in special cases where the motor has to fit into a narrow space.

It is emphasized that $\sigma$ and the TRV are being used here only as a means of producing an approximate size to form a basis for detailed design.

## 2.2 Choice of gear ratio in geared drives

Motors often require gearboxes to drive the load. The specific torque of a normal motor is quite small compared to the equivalent torque per unit volume available in mechanical or hydraulic drives. A gearbox is the obvious way to step up the motor torque $T_m$ by the required amount. If the gear ratio is $n$, the torque applied to the load is $nT_m$. The motor speed is increased over the load speed by the same ratio. In most cases the increased motor speed falls in a standard speed range for 'high-speed' motors, which may be typically anywhere from 2000 to 10 000 r.p.m.

# CHOICE OF GEAR RATIO IN GEARED DRIVES

If the gearbox efficiency is 100 per cent, the output power of the motor is equal to the power applied to the load. The choice of gear ratio depends on how the drive operates. If the speed is constant it is usually a simple matter of matching the load torque $T_L$ to the rated continuous motor torque $T_{mc}$:

$$n = \frac{T_L}{T_{mc}}.$$

If, however, the load has a 'dynamic' requirement which specifies a profile of speed or position as a function of time, the choice of the gear ratio and the motor parameters is more complicated.

## 2.2.1 Simple acceleration of pure inertia load

Referring to Fig. 2.3, if the motor torque is its peak rated torque $T_{mp}$, the acceleration of the load is given by

$$\alpha = \frac{T_{mp}}{n\left(J_m + \frac{J_L}{n^2}\right)}$$

where the term in brackets is the inertia of the motor combined with the load inertia referred to the motor shaft. If $n$ is large the gearing makes the load inertia insignificant, but reduces the load speed and acceleration relative to those of the motor. If $n$ is small the referred load inertia is large, and this limits the acceleration. Between the extremes of large and small $n$, there is a value that gives maximum acceleration for fixed values of $T_{mp}$ and the separate inertias. This 'optimum' value can be determined by equating the differential of the above expression to zero, giving

$$n = \sqrt{\frac{J_L}{J_m}}$$

which is a well-known result. This value of $n$ makes the referred load inertia

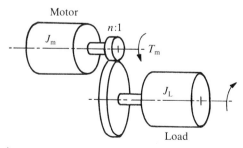

Fig. 2.3. Gear ratio.

26   SIZING, GEARING, AND TORQUE PRODUCTION

equal to the motor inertia. The maximum acceleration of the load is therefore

$$\alpha_{max} = \frac{1}{2} \frac{T_{mp}}{J_m} \frac{1}{n}.$$

The corresponding acceleration of the motor is $n$ times this value. In this analysis, the inertia of the gearbox has been ignored. For a very precise evaluation, in the case of a single-stage gearbox, the pinion inertias can be combined with (added to) the respective motor and load inertias.

### 2.2.2 Acceleration of inertia with fixed load torque

A slightly more complicated example is where the load has a fixed torque $T_L$ in addition to its inertia. This diminishes the torque available for acceleration, so the load acceleration is now given by

$$\alpha = \frac{T_{mp} - \dfrac{T_L}{n}}{n\left(J_m + \dfrac{J_L}{n^2}\right)}.$$

Again there is one value of gear ratio $n$ that produces maximum acceleration, and by the same differentiation process it is found to be

$$n = \frac{T_L}{T_{mp}}\left[1 + \sqrt{1 + \frac{J_L}{J_m} \cdot \frac{T_{mp}^2}{T_L^2}}\right].$$

If the inertias are unchanged from the previous case, the gear ratio is increased. The expression for the optimum ratio can be substituted back in the formula for acceleration to find the maximum load acceleration. The result is

$$\alpha_{max} = \frac{1}{2} \frac{T_{mp}}{J_m} \frac{1}{n}$$

which has exactly the same form as in the previous case; the difference is that with a larger ratio the load acceleration will be smaller. It is interesting to note that the maximum acceleration of the motor is unchanged, and is equal to one-half the torque/inertia ratio of the motor.

### 2.2.3 Peak/continuous torque ratio of motor

In the constant-speed case, the choice of $n$ maximizes the utilization of the continuous torque rating of the motor, $T_{mc}$. In the acceleration case, the choice of $n$ maximizes the utilization of the motor's peak acceleration capability as expressed by its peak torque/inertia ratio $T_{mp}/J_m$. Consider a load that requires both short periods of acceleration and long periods at constant speed. Then there is the question, can the two values of $n$ be the same? If so, the utilization of both aspects of motor capability will be maximized at the same time.

This problem can be solved analytically in a few special cases, and one solution is given here as an example of the kind of analysis that is needed to get a highly optimized system design. Assume that the load torque is constant at all times, but that short bursts of acceleration (or deceleration) are required from time to time. The peak rated torque of the motor will be used for acceleration, and the continuous rated torque for constant speed. If we equate the two separate values of $n$ from the appropriate formulas given above, and if we write

$$T_{mp} = kT_{mc}$$

where $k$ is the peak/continuous torque ratio of the motor, then the following relationship can be derived:

$$\frac{n^2 J_m}{J_L} = \frac{k^2}{(k-1)^2 - 1}.$$

The left-hand expression is the ratio of the referred motor inertia to the load inertia, and we can refer to it as the 'referred inertia ratio' or just 'inertia ratio'. For a range of values of the inertia ratio, the equation can be solved to find the values of $k$ that simultaneously optimize $n$ for both the constant-speed and acceleration periods. The most interesting result of this is that a large range of inertia ratio is encompassed by only a small range of values of $k$: as the inertia ratio changes from infinity down to 2, $k$ changes only from 2 to 4. But values of $k$ in this range are extremely common: so common, in fact, as to appear to be a natural characteristic of electric motors. This implies that for most inertia ratios where the referred motor inertia is more than twice the load inertia, the gear ratio can be chosen to make good utilization of both the continuous torque and the peak acceleration of the motor, provided $k \geqslant 2$. If $k < 2$, the gear ratio must be chosen for constant speed or for acceleration, and cannot be optimal for both.

## 2.2.4 *General speed and position profiles*

The cases considered are all idealized by rather restrictive assumptions that may be too simple in a complex motion-control system. For more detailed work it is sometimes useful to simulate the performance of the whole system, motor plus load. Many numerical experiments can then be performed to find out the best gear ratio and motor parameters. There are specialized software packages like Control-C, ACSL, and SIMNON to perform these simulations, and in principle they can deal with almost any degree of complexity in the speed/position profiles and indeed with multiple motor drives.

In conclusion it may be said that the property of electric motors to provide short bursts of peak torque for acceleration is one of the most important aspects of their use in motion control systems.

## 2.3 Basic principles of torque production

Consider a machine having one winding on the rotor and one on the stator. Let the self-inductances of the stator and rotor windings be $L_1$ and $L_2$ respectively, and let the mutual inductance between them be $L_{12}$. Let $\theta$ be the angular position of the rotor relative to some arbitrary reference position.

The total torque is given by

$$T = \frac{1}{2} i_1^2 \frac{dL_1}{d\theta} + \frac{1}{2} i_2^2 \frac{dL_2}{d\theta} + i_1 i_2 \frac{dL_{12}}{d\theta} \text{ Nm}.$$

Two kinds of torque can be identified: alignment torque and reluctance torque. The first two terms are reluctance torque and the third one is alignment torque. Any of these components is easy enough to visualize when it is the only component present.

Consider reluctance torque first, and refer to Fig. 2.4(a). If winding 1 is excited the rotor will tend to take up a position which minimizes the reluctance (and maximizes the inductance) of winding 1. The rotor becomes magnetized by induction and its induced magnetic poles tend to align with the excited

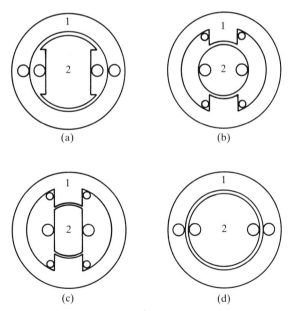

FIG. 2.4. Illustrating reluctance and alignment torques. (a) Reluctance torque only when winding 1 is excited. Alignment torque when both windings are excited. (b) Reluctance torque only when winding 2 is excited. Alignment torque when both windings are excited. (c) Reluctance torque when either winding is excited. Alignment torque when both windings are excited. (d) No reluctance torque. Alignment torque when both windings are excited.

poles of opposite polarity on the stator bore. If only winding 2 is excited there is no reluctance torque, because the reluctance and inductance of winding 2 do not depend on the rotor position.

In Fig. 2.4(b) the roles of rotor and stator are exchanged. Reluctance torque is produced if winding 2 is excited, but not winding 1. In Fig. 2.4(c) reluctance torque is produced if either winding is excited. In Fig. 2.4(d) there is no reluctance torque, because the inductance of both windings is independent of the rotor position.

The existence of reluctance torque depends on the saliency. In Figs 2.4(a) and (b) the machine is 'singly salient' and there is only one possible component of reluctance torque. In Fig. 2.4(c) the machine is 'doubly salient', meaning that there are poles or projections on both rotor and stator. Correspondingly there are two possible components of reluctance torque. In Fig. 2.4(d) there is no saliency and no reluctance torque; this is called a 'cylindrical' or 'non-salient pole' machine.

Alignment torque is produced in all four machines if both windings are excited. In Fig. 2.4(d) it is the only possible torque component. Alignment torque is caused by the tendency of excited rotor poles to align with excited stator poles, and requires both windings to be excited. In this case we say that the machine is 'doubly excited'. Reluctance torque, on the other hand, is possible in 'singly excited' machines, i.e. those with only one winding excited. It is characteristic of reluctance torque that excited poles on one member tend to align with induced poles on the other member.

In singly-excited machines all the magnetizing ampere-turns must be provided by the one winding, which is supplied with all the electric power. There is a conflict between the torque production and the minimization of these magnetizing ampere-turns in reluctance machines. The conflict shows up as a poor power factor in a.c. reluctance machines, and this in turn reduces the efficiency. In switched reluctance motors it may show up as a high converter volt-ampere requirement, or as high ripple current in the d.c. supply. The problem is less than in the synchronous reluctance motor, partly because of the torque-multiplying 'vernier' principle in the switched motor, which compensates the ratio of power output to converter volt-amperes, albeit at the expense of a higher operating frequency.

### 2.3.1 Production of smooth torque

Although both alignment and reluctance torques may be present together, they cannot be freely combined. The main restriction arises from the need to produce smooth, ripple-free torque. Consider each of the four machines of Fig. 2.4 in turn.

In (a), continuous torque can be produced if winding 1 is a polyphase winding with sinewave currents so phased as produce a rotating m.m.f. wave with corresponding excited poles on the stator bore that rotate at 'synchronous speed'. The rotor will align with these poles. If the rotor winding is excited, it must be with fixed d.c. excitation or magnets: its excited poles will then tend to align with the rotating stator poles. If the rotor is unexcited, it will have induced poles that tend to align with the rotating stator poles. In general the reluctance and alignment torques are both present. The sign of each of these torque components depends on the angular displacement between the rotor axis and the rotating m.m.f. wave. For maximum torque per ampere, obviously this angle must be in a range where they add. If the magnitudes of the rotating stator m.m.f. wave and the rotor excitation are constant, the torque will be constant. This machine is a polyphase a.c. synchronous machine with salient (rotor) poles.

In (b) the situation is similar but the roles of the windings are exchanged. Now the stator must carry fixed d.c. excitation or magnets. If the rotor carries a polyphase a.c. winding with sinewave currents, it will produce an m.m.f. wave that rotates relative to the rotor; if the rotor rotates physically at the same speed in the opposite direction, the rotor m.m.f. wave and its excited poles will be stationary, that is, 'synchronous' with the stationary stator poles. If the magnitudes of the rotor m.m.f. wave and the stator excitation are constant, the torque will be constant. This machine is an 'inside-out' or 'inverted' salient-pole synchronous machine. Of the two possible forms of synchronous machine, (a) is common and (b) is rare. The electric power converted into mechanical work is introduced through the polyphase winding, and since this power is far greater than that required by the d.c. winding, it is much easier to supply and cool the polyphase winding if it is stationary.

The machine of Fig. 2.4(b) is, however, extremely common in the form of the d.c. commutator motor. Here the stator has fixed d.c. excitation or magnets, as before. The rotor winding is a kind of polyphase winding with a large number of phases, the terminals of which are all brought out to the commutator. As the rotor rotates, the brushes bearing on the commutator repeatedly disconnect and reconnect the d.c. supply to the winding in such a way that the distribution of currents around the rotor surface remains fixed in space, and therefore 'synchronous with' or stationary with respect to the stator. By this means, alignment torque is produced. Again, if the currents in both members are constant, the torque is constant. In wound-field machines (but not PM machines) a reluctance torque may also be produced between the rotor current and the salient-pole stator structure, but for this it is necessary to rotate the axis on which the brushes are located, and this produces sparking which is highly undesirable. Indeed, d.c. machine designers go to some lengths to prevent sparking (by adding interpoles and compensating windings), and in so doing suppress the reluctance torque component; perhaps for this reason, its existence is rarely recognized.

Until relatively recently, doubly-salient machines, of which Fig. 2.4(c) is a simple example, were very rare; it is only with the growth of power electronics that they have been developed for useful applications, mainly as stepper motors. It is not possible with this structure to generate a pure rotating m.m.f. wave on either member, and for all practical purposes it is impossible to generate a smooth alignment torque with reasonable current waveforms in the two windings. The doubly-salient machine usually is a singly-excited reluctance motor, with only a stator winding. The current to this winding must be switched so that it flows only when the rotor poles are approaching alignment with the stator poles. When the poles are separating, the current must be substantially zero, otherwise a negative (braking) torque would be produced. The torque is therefore inherently pulsed, and the design requires special consideration (see Chapter 7). Unlike the other machines, there is no inherently simple way to make the torque constant and ripple-free.

In principle the doubly-salient machine could be rotor-fed rather than stator-fed, but this would require slip-rings and the rotor winding would be difficult to cool; it would also complicate the construction because of the need to retain the rotor winding against centrifugal force. When the windings are on the stator, however, the motor becomes very simple in construction and this is one of its main attractions.

As distinct from PM stepper motors (doubly-excited machines conforming to (b) or (d) with magnets on the rotor) and VR stepper motors (type (c)), the hybrid stepper motor is, in a sense, doubly excited because it has a multiple-phase stator winding as well as a rotor magnet. Its torque is almost entirely reluctance torque since the flux-linkage of the stator phases does not vary substantially with rotor position; rather, it is a flux-switched or magnet-assisted reluctance motor (Kenjo 1985).

Finally, the machine of Fig. 2.4(d) can be fitted with any combination of polyphase and d.c. windings on the stator and the rotor. With a polyphase a.c. winding on the stator and a d.c. winding on the rotor, the machine is a 'nonsalient pole' synchronous machine; it has alignment torque but no reluctance torque. With polyphase a.c. windings on both rotor and stator, the machine represents the common induction motor or any one of a large number of doubly-fed a.c. machines such as the Scherbius or Leblanc machines. In the case of the induction motor, the polyphase rotor currents are induced by slip. Again, in all these variants if the currents are constant d.c. or polyphase a.c., according to the type of winding, the torque is constant and essentially ripple-free.

In any of these machines a plain d.c. winding can be replaced by a permanent magnet. If the winding is a rotor winding, this eliminates the need for slip-rings (or a brushless exciter). Although the magnet excitation is fixed, this is a small price to pay for the convenience and simplicity provided by permanent magnets, particularly in smaller motors. Almost all small d.c. commutator motors have PM stators. The permanent-magnet synchronous machine is less common, possibly because synchronous machines are less common anyway, especially in small sizes where permanent magnets are practical.

# SIZING, GEARING, AND TORQUE PRODUCTION

The brushless d.c. PM motor is of the type shown in Fig. 2.4(d). In configuration it is identical to the non-salient pole synchronous machine, and indeed with a.c. polyphase windings on the stator it can operate as such, if the phase currents are sinusoidal and the windings are approximately sinusoidally distributed (see Chapter 5). It is more common, however, for the currents to be switched d.c. (see Chapter 4). In this form the brushless d.c. motor is truly an inside-out d.c. commutator motor with the mechanical commutator replaced by an electronic switching converter. Because of the cost of switches, the number of phases is kept much smaller than the equivalent number on the d.c. commutator motor rotor, and three is the commonest number used. With switched d.c. in this three-phase winding, the stator m.m.f. wave does not rotate at constant speed, and this is an important difference between the brushless d.c. motor and the a.c. PM synchronous motor. The production of smooth, ripple-free alignment torque is still theoretically possible, but is in practice not as good as with a true rotating-field machine.

Because of the small phase number in the brushless d.c. motor (usually three) the electronic commutation is quite coarse compared to that of a mechanical commutator, which usually has at least 12 segments and sometimes several hundred.

In view of the apparent simplicity of a.c. motors, brushless d.c. motors, and even the doubly-salient reluctance motor, it may seem surprising that so many adjustable-speed drives use d.c. commutator motors. The basic reason for this is that the speed of the commutator motor is proportional to the voltage, whereas the speed of all the other motors is proportional to the fundamental frequency in the a.c. (or switched d.c.) windings; and electronic voltage controllers are less expensive and less complex than electronic frequency controllers.

## Problems for Chapter 2

1. Calculate the rotor diameter of a motor that develops 100 W shaft power at 1500 r.p.m. with a torque per unit rotor volume of 12 kN m/m$^3$ and a rotor length/diameter ratio of 1.4.

2. An a.c. motor has an electric loading of 12 A/mm and an average airgap flux-density of 0.6 T. What is the torque per unit rotor volume and the electromagnetic shear stress in the airgap? Assume a fundamental winding factor of 0.9, and assume that the rotor flux is orthogonal to the stator ampere-conductor distribution.

3. In the motor of Problem 2, the slot fill factor is 0.42, the slot depth is 18 mm, and the tooth-width to tooth-pitch ratio is 0.56. Estimate the current density in the winding.

4. A brushless servo motor has a peak rated torque of 14 N m and a torque/inertia ratio of 35 000 rad/s$^2$. It is found to be capable of accelerating an inertial load at 4600 rad/s$^2$ through a 2:1 speed-reduction gearbox. Estimate the inertia of the load in kg m$^2$. What value of gear ratio would result in the maximum acceleration of this motor and load combination, and what would be the value of the maximum acceleration?

5. A limited-rotation actuator has a rotor winding and a stator winding. The geometry is such that the self-inductances are constant but the mutual inductance varies from zero to 109 mH in a rotation of 75°. Calculate the torque when the windings are connected in series carrying a current of 10 A.

# 3 Permanent-magnet materials and circuits

## 3.1 Permanent-magnet materials and characteristics

The sustained success of the permanent-magnet industry in developing improved magnet characteristics is evident from Fig. 3.1, the latest addition being neodymium-iron-boron which has been pioneered by Sumitomo as 'Neomax', General Motors as 'Magnequench', Crucible ('Crumax'), and IG Technologies ('NeIGT'). At room temperature NdFeB has the highest energy product of all commercially available magnets. The high remanence and coercivity permit marked reductions in motor framesize for the same output compared with motors using ferrite (ceramic) magnets. However, ceramic magnets are considerably cheaper.

Both ceramic and NdFeB magnets are sensitive to temperature and special care must be taken in design for working temperatures above 100°C. For very high temperature applications Alnico or rare-earth cobalt magnets must be used, for example 2–17 cobalt–samarium which is useable up to 200°C or even 250°C.

NdFeB is produced either by a mill-and-sinter process (Neomax) or by a melt-spin casting process similar to that used for amorphous alloys (Magnequench). Powder from crushed ribbon is bonded or sintered to form

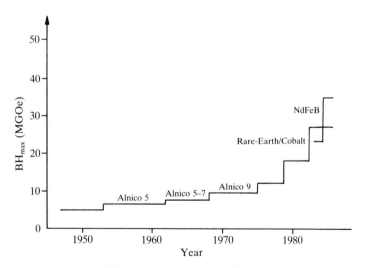

FIG. 3.1. Development of PM materials in terms of maximum energy product.

# PERMANENT-MAGNET MATERIALS AND CHARACTERISTICS

the MQI or MQII grades produced by Magnequench Division of GM. The MQI bonded magnets can be formed in a wide variety of shapes. They are not 100 per cent dense and coatings may be used to prevent corrosion. With MQII and other sintered materials a dichromate coating may be used, or electroplating.

For lowest cost, ferrite or ceramic magnets are the universal choice. This class of magnet materials has been steadily improved and is now available with remanence of 0.38 T and almost straight demagnetization characteristic throughout the second quadrant. The temperature characteristics of ferrite magnets can be tailored to the application requirements so that maximum performance is obtained at the normal operating temperature, which may be as high as 100°C.

A brief summary of magnet properties is given in Table 3.1. More detail can be obtained from suppliers' data sheets, as the examples show. Specialist data and measurements are often made by permanent-magnet research and development bodies; for example, in the UK, the Magnet Centre at Sunderland Polytechnic, and in the USA, the University of Dayton, Ohio. Activity in magnet research is also well reported in IEEE and specialist conference proceedings.

**Table 3.1.** Magnet properties

| Property | Units | Alnico 5–7 | Ceramic | $Sm_2Co_{17}$ | NdFeB |
|---|---|---|---|---|---|
| $B_r$ | T | 1.35 | 0.405 | 1.06 | 1.12 |
| $\mu_0 H_c$ | T | 0.074 | 0.37 | 0.94 | 1.06 |
| $(BH)_{max}$ | MGOe | 7.5 | 3.84 | 26.0 | 30.0 |
| $\mu_{rec}$ | | 1.9 | 1.1 | 1.03 | 1.1 |
| Specific gravity | | 7.31 | 4.8 | 8.2 | 7.4 |
| Resistivity | $\mu\Omega$ cm | 47 | $>10^4$ | 86 | 150 |
| Thermal expansion | $10^{-6}/°C$ | 11.3 | 13 | 9 | 3.4 |
| $B_r$ temperature coefficient | %/°C | −0.02 | −0.2 | −0.025 | −0.1 |
| Saturation $H$ | kOe | 3.5 | 14.0 | >40 | >30 |

## 3.2 B-H loop and demagnetization characteristics

The starting-point for understanding magnet characteristics is the $B$–$H$ loop or 'hysteresis loop', Fig. 3.2. The x-axis measures the magnetizing force or 'field intensity' $H$ in the material. The y-axis is the magnetic flux-density $B$ in the material. An unmagnetized sample has $B=0$ and $H=0$ and therefore starts out at the origin. If it is subjected to a magnetic field, as for example in a magnetizing fixture (an electromagnet with specially shaped pole pieces to

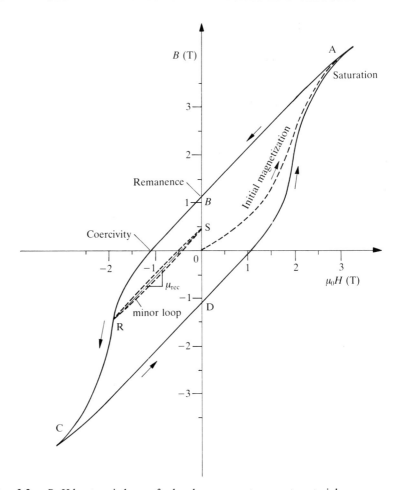

FIG. 3.2. B–H hysteresis loop of a hard permanent magnet material.

focus flux into the magnet), then $B$ and $H$ in the magnet will follow the curve OA as the external ampere-turns are increased. If the external ampere-turns are switched off, the magnet relaxes along AB. Its operating point $(H, B)$ will depend on the shape of the magnet and the permeance of the surrounding 'magnetic circuit'. If the magnet is surrounded by a highly permeable magnetic circuit, that is, if it is 'keepered', then its poles are effectively shorted together so that $H=0$ and the flux-density is then the value at point $B$, the remanence $B_r$. The remanence is the maximum flux-density that can be retained by the magnet at a specified temperature after being magnetized to saturation.

External ampere-turns applied in the opposite direction cause the magnet's operating point to follow the curve from B through the second quadrant to C, and again if they are switched off at C the magnet relaxes along CD. It is now magnetized in the opposite direction and the maximum flux-density it can retain when 'keepered' is $-B_r$. To bring the flux-density to zero from the original positive remanence the external ampere-turns must provide within the magnet a negative magnetizing force $-H_c$, called the coercivity. Likewise, to return the flux-density to zero from the negative remanence point D, the field $+H_c$ must be applied. The entire loop is usually symmetrical and can be measured using special instruments such as the Hysteresisgraph made by Walker Scientific Instruments.

If negative external ampere-turns are applied, starting from point B, and switched off at R, the operating point of the magnet 'recoils' along RS. If the magnet is still 'keepered' the operating point ends up at point S. Now if the external ampere-turns are re-applied in the negative direction between S and R, the operating point returns along SR. The line RS is actually a very thin 'minor hysteresis loop' but for practical purposes it can be taken as a straight line whose slope is equal to the recoil permeability. This is usually quoted as a relative permeability, so that the actual slope of RS is $\mu_{rec}\mu_0$ H/m. Operation along RS is stable provided that the operating point does not go beyond the boundary of the original hysteresis loop.

A 'hard' PM material is one in which the hysteresis loop is straight throughout the second quadrant, where the magnet normally operates in service. In this case the recoil line is coincident with the second-quadrant section of the hysteresis loop. This is characteristic of ceramic, rare-earth/cobalt, and NdFeB magnets, and the recoil permeability is usually between 1.0 and 1.1. 'Soft' PM materials are those with a 'knee' in the second quadrant, such as Alnico. While Alnico magnets have very high remanence and excellent mechanical and thermal properties, they are limited in the demagnetizing field they can withstand. It should be noted that compared with lamination steels even the 'soft' PM materials are very 'hard': in other words, the hysteresis loop of a typical nonoriented electrical steel is very narrow compared with that of even the Alnico magnets.

The most important part of the B-H loop is the second quadrant, drawn in more detail in Fig. 3.3. This is called the demagnetization curve. In the absence of externally applied ampere-turns the magnet's operating point is at the intersection of the demagnetization curve and the 'load line', whose slope is the product of $\mu_0$ and the permeance coefficient of the external circuit (see Chapter 4, Section 4.2).

Since B and H in the magnet both vary according to the external circuit permeance, it is natural to ask what it is about the magnet that is 'permanent'. The relationship between B and H in the magnet can be written

$$B = \mu_0 H + J.$$

The first term is the flux-density that would exist if the magnet were removed and the magnetizing force remained at the value H. Therefore the second term can be regarded as the contribution of the magnet to the flux-density within its

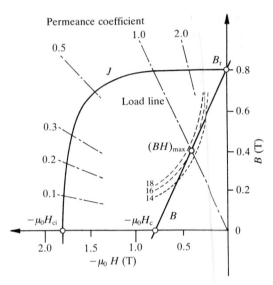

Fig. 3.3. Second-quadrant demagnetization characteristic showing intrinsic curve.

own volume. Clearly if the demagnetization curve is straight, and if its relative slope and therefore the recoil permeability are both unity, then $J$ is constant. This is shown in Fig. 3.3 for values of negative $H$ up to the coercivity. $J$ is called the magnetization of the magnet. Obviously it has the units of flux density, T. In most hard magnets the recoil permeability is slightly greater than 1 and there is a slight decrease of $J$ as the negative magnetizing force increases, but this is reversible down to the 'knee' of the $B$–$H$ loop (which may be in either the second or the third quadrant, depending on the material and its grade).

The magnetization and other parameters of the linear or 'recoil line' model are used in finite-element analysis for calculation of magnetic circuits, and several commercial packages are available to do this.

Evidently the magnet can recover or recoil back to its original flux-density as long as the magnetization is constant. The coercive force required to permanently demagnetize the magnet is called the intrinsic coercivity and this is shown as $H_{ci}$.

Another parameter often calculated is the magnet energy product, which is simply the product of $B$ and $H$ in the magnet. This is not the actual stored magnet energy, which depends on the history or trajectory by which the magnet arrived at its operating point and usually cannot be calculated except under very artificial conditions. The energy product is a measure of the stored energy but, more importantly, it gauges how hard the magnet is working to provide flux against the demagnetizing influence of the external circuit. Contours of constant energy product are rectangular hyperbolas and these are frequently drawn on graphical property data sheets provided by magnet suppliers. The maximum energy product or $(BH)_{max}$ of a given magnet occurs where the demagnetization characteristic is tangent to the hyperbola of its $(BH)_{max}$ value. If the recoil permability is unity, this occurs for a permeance coefficient of unity, provided that there are no externally applied ampere-turns from windings or other magnets.

Figure 3.4(a) shows a simple magnetic circuit in which the magnet is 'keepered' by a material or core of relative permeability $\mu_r$. Applying Ampere's law, and assuming uniform magnetizing force in both the magnet and the core,

$$H_m l_m + H_k l_k = 0.$$

This is effectively the line integral of $H$ around the magnetic circuit, and it is zero because there are no externally applied ampere-turns. Hence

$$H_m = -\frac{l_k}{l_m} H_k$$

which establishes that the magnet works in the second quadrant of the $B$–$H$ loop. Now consider the magnetic circuit of Fig. 3.4(b), in which the magnet flux is guided to a working space or airspace by infinitely permeable flux-guides. 'Electrical steel' is sufficiently permeable to satisfy this idealization. The above relationship becomes

$$H_m = -\frac{l_g}{l_m} H_g$$

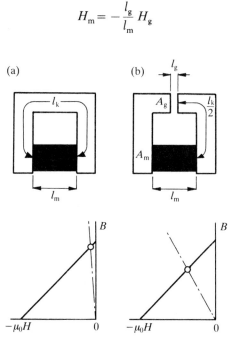

FIG. 3.4. Simple magnetic circuits illustrating the magnet operating point under different conditions, with no externally applied ampere turns.

where the subscript 'g' refers to the working airspace or airgap. By Gauss's law, the flux-densities in the magnet and the airgap are related by

$$B_m A_m = B_g A_g.$$

If we take the ratio of $B_m$ to $\mu_0 H_m$, recognizing that $B_g = \mu_0 H_g$,

$$\frac{B_m}{H_m} = -\mu_0 \frac{A_g}{A_m} \frac{l_m}{l_g} = -\mu_0 \mathrm{PC}$$

where PC is the permeance coefficient. The ratio of magnet pole area to airgap area is sometimes called the flux-concentration factor or flux-focusing factor:

$$C_\Phi = \frac{A_m}{A_g}.$$

A high permeance coefficient requires a low flux-concentration factor and/or a long magnet. Therefore, if the magnet pole area is made much larger than the airgap area in an attempt to raise the working flux-density in the airspace, the price paid is that the magnet works further down the demagnetization curve, and its margin against demagnetization is reduced. This can be compensated by making the magnet longer in the direction of magnetization.

The energy product is given by

$$B_m H_m = \frac{B_g H_g A_g l_g}{A_m l_m} = \frac{2 W_g}{V_m}$$

where $W_g$ is the magnetic energy stored in the airspace volume and $V_m$ is the volume of the magnet. Evidently the minimum magnet volume required to magnetize a given working volume of airspace is inversely proportional to the working energy product. Therefore, in these cases it pays to design the magnet length and pole area in such proportions relative to the length and area of the airspace, as to cause the magnet to work at $(BH)_{max}$. In motors this principle cannot be applied so simply, because the armature current produces demagnetizing ampere-turns that may be very great under fault conditions. To eliminate the risk of demagnetization, motors are designed so that on open-circuit or no-load the magnet operates at a high permeance coefficient (corresponding to a small airgap length) with adequate margin of coercive force to resist the maximum demagnetizing ampere-turns expected under load or fault conditions.

The remaining diagrams in Fig. 3.4 illustrate the relative intensities of $B$ and $H$ under different working conditions, in all cases with no externally applied ampere-turns. Note that $B$ is continuous (because it obeys Gauss's law), but $H$ is not. The discontinuities of $H$ are, of course, associated with the appearance of magnetic poles at the interfaces between different sections of the magnetic circuit, notably at the 'poles' of the magnet and the working airspace. The polarization of surfaces gives rise to a magnetic potential difference across the airspace which is useful for calculating flux distribution in motors. In Fig. 3.4(b) this potential difference is

$$u = H_g l_g \text{At}$$

with the units of ampere-turns. The corresponding magnetic potential difference across the magnet is

$$-u = H_m l_m \text{At}.$$

There is still widespread use of c.g.s. units in the magnet industry, whereas motors are usually designed in metric (SI) units in Europe and Japan, and in metric or mixed units in the USA. The most important conversion factors are shown in Table 3.2.

Table 3.2. Conversion factors for common units

| | |
|---|---|
| 1 inch | = 25.4 mm |
| 1 T | = 10.0 kG |
| 1 kA/m | = $4\pi$ Oe |
| 1 kJ/m$^3$ | = $\pi/25$ MGOe |

## 3.3 Temperature effects: reversible and irreversible losses

### 3.3.1 High-temperature effects

Exposure to sufficiently high temperatures for long enough periods produces metallurgical changes which may impair the ability of the material to be magnetized and may even render it non-magnetic. There is also a temperature, called the Curie temperature, at which all magnetization is reduced to zero. After a magnet has been raised above the Curie temperature it can be remagnetized to its prior condition provided that no metallurgical changes have taken place. The temperature at which significant metallurgical changes begin is lower than the Curie temperature in the case of the rare-earth/cobalt magnets, NdFeB, and Alnico; but in ceramic ferrite magnets it is the other way round. Therefore ceramic magnets can be safely demagnetized by heating them just above the Curie point for a short time. This is useful if it is required to demagnetize them for handling or finishing purposes. Table 3.3 shows these temperatures for some of the important magnets used in motors.

### 3.3.2 Reversible losses

The $B$–$H$ loop changes shape with temperature. Over a limited range the changes are reversible and approximately linear, so that temperature coefficients for the remanence and coercivity can be used. Sometimes a

**Table 3.3.** Metallurgical change and Curie temperature

|  | Metallurgical change (°C) | Curie temperature (°C) |
|---|---|---|
| Alnico 5 | 550 | 890 |
| Ceramics | 1080 | 450 |
| SmCo$_5$ | 300 | 700 |
| Sm$_2$Co$_{17}$ | 350 | 800 |
| NdFeB | 200 | 310 |

coefficient is also quoted for the flux density at the maximum-energy point. Table 3.4 gives some typical data. Ceramic magnets have a positive coefficient of $H_c$, whereas the high-energy magnets lose coercivity as temperature increases. In ceramic magnets the knee in the demagnetization curve moves down towards the third quadrant, and the permeance coefficient at the knee decreases. Thus ceramic magnets become better able to resist demagnetization as the temperature increases up to about 120°C. The greatest risk of demagnetization is at low temperatures when the remanent flux density is high and the coercivity is low; in a motor, this results in the highest short-circuit current when the magnet is least able to resist the demagnetizing ampere-turns. In high-energy magnets the knee moves the other way, often starting in the third quadrant at room temperature and making its way well into the second quadrant at 150°C. Grades with a high resistance to temperature are more expensive, yet these are often the ones that should be used in motors, particularly if high temperatures are possible (as they usually are under fault conditions).

All the magnets lose remanence as temperature increases. For a working temperature of 50°C above an ambient of 20°C, for instance, a ceramic magnet will have lost about 10 per cent. This is spontaneously recovered as the temperature falls back to ambient.

**Table 3.4.** Reversible temperature coefficients (%/°C)

|  | $B_r$ | $H_c$ | $H_{ci}$ |
|---|---|---|---|
| Alnico 5–7 | −0.02 | | |
| Ceramics | −0.19 | 0.20 | |
| SmCo$_5$ | −0.04 | | −0.25 |
| Sm$_2$Co$_{17}$ | −0.02 | | −0.20 |
| NdFeB | −0.11 | | −0.60 |

### 3.3.3 Irreversible losses recoverable by remagnetization

*(a) Domain relaxation.* Immediately after magnetization there is a very slow relaxation, starting with the least stable domains returning to a state of lower potential energy. The relaxation rate depends on the operating point and is worse below $(BH)_{max}$, i.e. at low permeance coefficients. In modern high-coercivity magnets at normal temperatures this process is usually negligible, particularly if the magnets have been stabilized (by temperature cycling and/or a.c. flux reduction) immediately after magnetization. Elevated temperatures during subsequent operation may, however, cause an increased relaxation rate. This can be prevented by temperature-cycling in the final assembly over a temperature range slightly wider than the worst-case operating range. Subsequent relaxation is reduced to negligible levels by this means. Figure 3.5 shows the 'natural' stability of different magnet materials at 24°C.

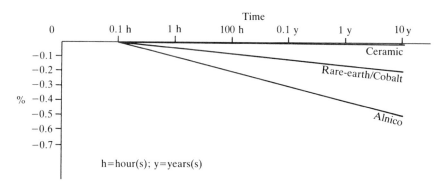

FIG. 3.5. Long-term stability of magnets at 24°C.

*(b) Operating point effect.* Temperature alters the B-H loop. If this causes the operating point to 'fall off' the lower end of a recoil line, there will be an irreversible flux loss. This is illustrated in Fig. 3.6. Operation is initially at point a on the load line 0a, which is assumed to remain fixed. The remanent flux-density corresponding to point a is at point A. When the temperature is raised from $T_1$ to $T_2$ the operating point moves from a to b, and the corresponding remanent flux-density moves from A to B'. Note that because the knee of the curve has risen above point b, the effective remanent flux-density is at B and is less than that at B', which is what it would have been if the magnet had been working at a high permeance coefficient.

If the temperature is now reduced to $T_1$ the operating point can recover only to a', which lies on the recoil line through A'. There has been a reversible

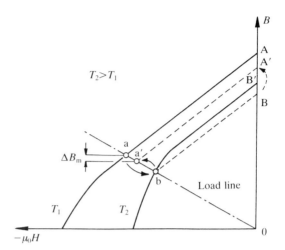

FIG 3.6. Irreversible loss caused by reversible changes in the $B$–$H$ loop.

recovery of remanence from B′ to A′, but not to A. The magnet has thus suffered an irreversible loss that can be recovered only by remagnetization at the lower temperature. If the whole cycle of changes is repeated it stabilizes with the remanence at A′ at the lower temperature $T_1$.

Manufacturers' data for irreversible loss should be interpreted carefully to distinguish between the long-term stability and the effects just described. Since the irreversible loss is dependent on the conditions of the application, in particular the permeance coefficient, irreversible loss is usually quoted at a fixed permeance coefficient. If the magnet is used at a higher permeance coefficient, the irreversible loss over the same temperature range will be lower.

### 3.4 Mechanical properties, handling, and magnetization

Magnets are often brittle and prone to chipping, but proper handling procedures are straightforward enough as long as the rules are followed. Modern high-energy magnets are usually shipped in the magnetized condition, and care must be taken in handling to avoid injury that may be caused by trapped fingers. A further hazard is that when two or more magnets are brought close together they may flip and jump, with consequent risk to eyes. Table 3.5 summarizes some of the important safety precautions.

The best way to 'tame' magnetized magnets is to keeper them. Fixtures for inserting magnets can be designed so that the magnets slide along between steel guides which are magnetically short-circuited together. There still remains the problem of entering the magnets between the guides, but usually

## MECHANICAL PROPERTIES

**Table 3.5.** Magnet safety

---

**Permanent magnets require strict adherence to safety procedures at all stages of handling and assembly**

1. **Always wear safety glasses when handling magnets.** This is particular important when assembling magnets into a motor. When a large pole magnet is being assembled from smaller magnets, the magnets have a tendency to flip and jump unexpectedly and may fly a considerable distance

2. **Work behind a plexiglass screen when experimenting or assembling magnet assemblies.** Watch out for trapped fingers, especially with large magnets or high-energy magnets

3. **Avoid chipping** by impact with hard materials, tools, or other magnets

4. **Never dry-grind rare-earth magnets**—the powder is combustible. In case of fire, use LP argon or nitrogen dry chemical extinguishers—never use water or halogens

5. **PM motors generate voltage when the shaft is rotated, even when disconnected from the supply.** This may be obvious to an engineer, but is a potential safety hazard for electricians and maintenance personnel. Use suitable warning labels, especially on large machines

6. **When assembling a rotor to a stator, with either one magnetized, the rotor must be firmly guided and the stator firmly located.** Never leave magnetized members open or unprotected

---

there is enough space to allow this to be done gently. Obviously it is important to keep magnets clear of watches and electronic equipment sensitive to magnetic fields. Floppy disks and magnetic tapes are particularly vulnerable, and high-energy magnets can seriously distort the image on computer terminals and monitors.

Magnets are usually held in place by bonding (for example with Loctite Multibond) or compression clips. In motors with magnets on the rotor, adhesive bonding is adequate for low peripheral speeds and moderate temperatures, but for high speeds a Kevlar banding or stainless steel retaining shell can be used. In motors it is not advisable to make the magnet an integral part of the structure. Mechanically, the magnet should be regarded as a 'passenger' for which space and fixturing must be provided. The important requirements are that the magnet should not move and that it should be protected from excessive temperatures.

A very wide range of shapes is possible, but in motors the most common are arcs and sometimes rectangles. Tolerances in the magnetized direction can be held very close, $\pm 0.1$ mm even for standard magnets. If the design permits a relaxation of the required tolerance, particularly in the dimensions perpendicular to the magnetic axis, this should be exploited because it reduces the cost of the finished magnets.

Thermal expansion of magnets is usually different in the directions parallel and perpendicular to the magnetic axis. The coefficients in Table 3.1 are for the direction of magnetization, i.e. along the magnetic axis. Most magnets have a high compressive strength but should never be used in tension or bending.

Magnetization of high-energy magnets requires such a high magnetizing force that special fixtures and power supplies are essential, and this is one reason why high-energy magnets are usually magnetized before shipping. The magnetizing force $H$ must be raised at least to the saturation level shown in Table 3.1, and this normally requires ampere-turns beyond the steady-state thermal capability of copper coils. Therefore, pulse techniques are used, or in some cases superconducting coils. Richardson (1987) discusses a technique for focusing the m.m.f. of pulsed windings into high-energy magnets.

Ceramic and Alnico magnets can sometimes be magnetized *in situ* in the final assembly, but this is almost never possible with high-energy magnets.

## 3.5 Application of permanent magnets in motors

Permanent magnets provide a motor with life-long excitation. The only outlay is the initial cost, which is reflected in the price of the motor. It ranges from a few pennies for small ferrite motors, to several pounds for rare-earth motors.

Broadly speaking, the primary determinants of magnet cost are the torque per unit volume of the motor, the operating temperature range, and the severity of the operational duty of the magnet.

### 3.5.1 Power density

For maximum power density the product of the electric and magnetic loadings of the motor must be as high as possible. The electric loading is limited by thermal factors, and also by the demagnetizing effect on the magnet. A high electric loading necessitates a long magnet length in the direction of magnetization, to prevent demagnetization. It also requires a high coercivity, and this may lead to the more expensive grades of material (such as 2–17 cobalt–samarium), especially if high temperatures will be encountered.

The magnetic loading, or airgap flux, is directly proportional to the remanent flux density of the magnet, and is nearly proportional to its pole face area. A high power density therefore requires the largest possible magnet volume (length times pole area).

With ceramic magnets the limit on the magnet volume is often the geometrical limit on the volume of the rotor itself, and the highest power densities cannot be obtained with these magnets. With rare-earth or other high-energy magnets, the cost of the magnet may be the limiting factor.

The airgap flux-density of a.c. motors is limited by saturation of the stator teeth. Excessive saturation absorbs too much excitation m.m.f. (requiring a disproportionate increase in magnet volume) or causes excessive heating due to core losses. For this reason there is an upper limit to the useable energy of a permanent magnet. With a straight demagnetization characteristic throughout the second quadrant and a recoil permeability of unity, the maximum energy-product $(BH)_{max}$ is given by

$$(BH)_{max} = \frac{B_r^2}{4\mu_0} \text{ J/m}^3.$$

Assuming that the stator teeth saturate at 1.8 T and that the tooth width is half the tooth pitch, the maximum airgap flux-density cannot be much above 0.9 T and is usually lower than this. Therefore there will be little to gain from a magnet with a remanent flux-density above about 1 or 1.2 T, implying that the highest useable energy product is about 300 kJ/m³ (equivalent to 35–40 MGOe). At 100°C, such characteristics are barely within the range of the best available high-energy magnets. According to this argument, it is just as important to develop magnetic materials with 'moderate' properties and low cost, as it is to develop 'super magnets' regardless of cost. The long-awaited material with cobalt–samarium properties at ceramic prices is unfortunately still awaited, although progress is being made with neodymium–iron–boron.

### 3.5.2 Operating temperature range

Figure 3.7 shows demagnetization curves for several common motor magnets at (a) 25 and (b) 125°C. Because of the degradation in the remanent flux density and in the coercive force, the choice of material and the magnet volume must usually be determined with reference to the highest operating temperature. Fortunately brushless motors have very low rotor losses. The stator is easily cooled because of the fine slot structure and the proximity of the outside air. Consequently the magnet can run fairly cool (often below 100°C) and it is further protected by its own thermal mass and that of the rest of the motor. The short-time thermal overload capability of the electronic controller would normally be less than that of the motor, providing a further margin of protection against magnet overtemperature.

### 3.5.3 Severity of operational duty

Magnets can be demagnetized by fault currents such as short-circuit currents produced by inverter faults. In brushless motors with electronic control the problem is generally limited by the protective measures taken in the inverter and the control. With an over-running load, or where two motors are coupled to a single load, shorted turns or windings can be troublesome because of drag torque and potential overheating of the stator. But by the same token, the dynamic braking is usually excellent with a short-circuit applied to the motor

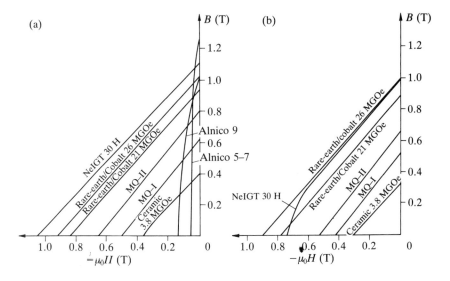
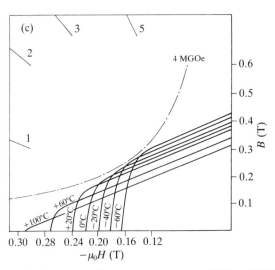

FIG. 3.7. Demagnetization curves for motor magnets at 25°C and 125°C. (a) and (b) Common motor magnets; (c) Temperature behaviour of a high-energy ceramic magnet; (d) Manufacturer's data for a Neodymium–Iron–Boron magnet.

# APPLICATION OF PERMANENT MAGNETS IN MOTORS 49

(d)

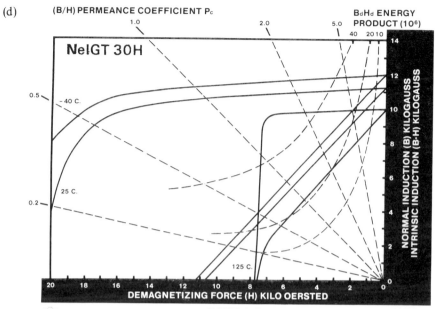

**DEMAGNETIZATION CURVES**

I G Technologies, Inc.
405 Elm Street
Valparaiso, Indiana 46383 U.S.A.
(219) 462-3131

| Magnetic Characteristics | | NeIGT 30H |
|---|---|---|
| Residual Induction - Br | (kG) | 11.2 |
| | (mT) | 1120 |
| Coercive Force - Hc | (kOe) | 10.6 |
| | (kA/m) | 810 |
| Intrinsic Coercive Force - Hci | (kOe) | >17.0 |
| | (kA/m) | >1275 |
| Peak Energy Density (BdHd) max | (MGOe) | 30 |
| | (KJ/m³) | 240 |
| Magnetizing Force - Hs | (kOe) | >30.0 |
| | (kA/m) | >2765 |
| Hk (H @ .9 Br) | (kOe) | 14.3 |
| Recoil Permeability | (ΔB/ΔH) | 1.1 |
| Reversible Temperature Coefficient for Bd (%/C.) | | .10 |
| Magnetic Orientation* | | T |
| **Material Characteristics** | | |
| Density | (lb/in³) | .268 |
| | (g/cm³) | 7.4 |
| Curie Temperature (C.) | | 310 |
| Temperature Affecting Material (C.) | | 200 |
| Rockwell Hardness R | | C58 |
| Electrical Resistivity of Material (at 25 C.) | (microohm cm) | 150 |
| Coefficient of Thermal Expansion (x 10⁻⁶/C.) | (E //M) | 3 |
| | (E⊥M) | -5 |
| Tensile Strength | (10³ psi) | 11.5 |

\* U - Unoriented   A - Axial
  T - Transverse   I - Isostatic

Relative Br vs. Magnetizing Field H for NeIGT

*Br after magnetizing in 50,000 Oe field is taken as the 100% value.

Virgin and Previously Magnetized material are the same.

(d)

terminals, and motors may well be designed to take advantage of this. As is often the case, characteristics that are desirable for one application are undesirable for another. The design must accommodate all the factors that stress the magnet, not only electromagnetic but thermal and mechanical as well.

## Problems for Chapter 3

1. Sketch the complete $B-H$ curve for a typical 'hard' permanent magnet. Indicate the remanent flux-density and the coercive force.

2. In which of the following electrical machines would you expect to find permanent magnets?
    (a) induction motor;
    (b) brushless d.c. servomotor;
    (c) variable-reluctance stepper motor.

3. Which of the following motors never has permanent magnets?
    (a) d.c. commutator motor;
    (b) induction motor.

4. Which of the following materials is a 'hard' permanent magnet?
    (a) samarium cobalt;
    (b) stainless steel;
    (c) gallium arsenide.

5. In the following motors, are the permanent magnets on the rotor or the stator?
    (a) d.c. commutator motor;
    (b) brushless d.c. motor.

6. State Ampere's Law.

7. Which of the following units are identical to the henry?
    (a) Wb/A t (Webers per ampere-turn)
    (b) A t/Wb (Ampere-turns per Weber)
    (c) V s/A (Volt-seconds per ampere)

8. Draw the demagnetization characteristic of a typical hard permanent magnet and indicate the recoil permeability.

9. Define the energy product of a permanent magnet.

10. If magnetic flux could be bottled, how much energy would be stored in a one-litre bottle containing flux at a uniform density of 1.0 T?

11. Large magnets are sometimes assembled from smaller magnets as a wall is built from bricks. It is necessary to test the polarity of each small magnet before adding it to the assembly. If the material is very high energy neodymium–iron–boron, which of the following polarity tests would you recommend?
    (a) floating the small magnet on a cork in water and noting which way it points in the earth's field;
    (b) offering the small magnet up to a compass and noting which way the compass needle swings;

(c) offering the small magnet up to another small magnet whose polarity is known, and noting whether the poles attract or repel.

Give reasons for your choice.

12. Figure 3.8 shows the demagnetization characteristic of a permanent magnet and the load line of a magnetic circuit in which it is used. Estimate graphically the percentage loss in remanent flux-density if the temperature is raised to 125°C and maintained there (A'). Assume that the initial operating point (A) is at 25°C.

13. Figure 3.9 shows a magnet used in magnetic resonance (MR) experiments for noninvasive medical diagnosis. Calculate the flux-density in the work space if the

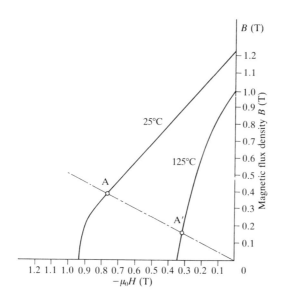

FIG. 3.8.

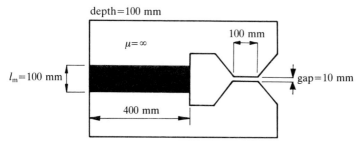

FIG. 3.9.

permanent magnet is ferrite with a remanent flux-density of 0.4 T and a straight demagnetization characteristic with recoil permeability of 1.0.

14. Figure 3.10 shows a rotary device excited by a permanent magnet whose demagnetization curve is straight throughout the second quadrant of the hysteresis loop. The remanent flux density is 1.0 T and the relative recoil permeability is 1.0. Calculate the airgap flux-density if the magnet length is 18 mm, assuming that the rotor is in the 'aligned' position as shown.

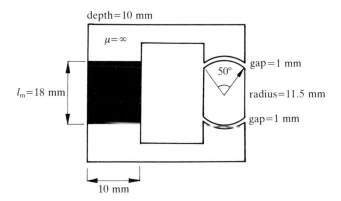

FIG. 3.10.

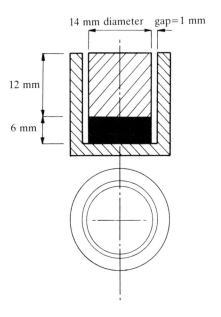

FIG. 3.11.

15. For the conditions of Problem 14, determine the values of the magnetizing force $H$ in the magnet and the total m.m.f. across it. Draw the second quadrant demagnetization characteristic and indicate the magnet operating point on it.

16. Figure 3.11 shows a loudspeaker magnet assembly. If the magnet has a straight demagnetization characteristic with a remanent flux-density of 1.0 T and a relative recoil permeability of 1.0, calculate the magnetic flux-density in the airgap. Neglect all fringing and leakage effects, and assume that the steel parts are infinitely permeable. Draw a diagram showing the flux path. What is the maximum energy product of the magnet? What is its actual energy product when it is installed in the assembly?

# 4 Squarewave permanent-magnet brushless motor drives

## 4.1 Why brushless d.c.?

The brushless d.c. motor is shown in its most usual form in Figure 4.1 alongside the PM d.c. commutator motor, and Figures 4.2(a)–(d) show examples of motors in commercial production. The stator structure is similar to that of a polyphase a.c. induction motor. The function of the magnet is the same in both the brushless motor and the d.c. commutator motor. In both cases the airgap flux is ideally fixed by the magnet and little affected by armature current.

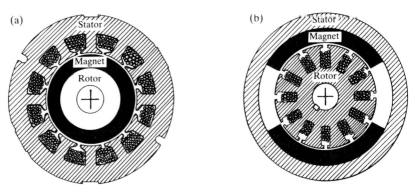

FIG. 4.1. (a) Brushless d.c. motor and (b) PM d.c. commutator motor.

The most obvious advantage of the brushless configuration is the removal of the brushes. Brush maintenance is no longer required, and many problems associated with brushes are eliminated. For example, brushes tend to produce RFI (radio-frequency interference) and the sparking associated with them is a potential source of ignition in inflammable atmospheres. These problems should not be overstated, however. RFI at least has the advantage of high frequency, so that filter components need not be very large. This is not necessarily the case with the lower-order harmonics associated with the commutation of the brushless motor. Commutator motors are quite commonly immersed in automobile petrol tanks to drive the fuel pump. This shows that they are not automatically excluded from 'hazardous' environments.

The problems that arise with commutator motors are sometimes not so obvious. In some applications the accumulation of brush debris or dust is a problem, particularly if it gets into the bearings or if it forms a conducting track that leads to flashover. The operation and life of brushes depend on factors such as atmospheric conditions, which may necessitate the use of different brush grades in the same motor operating in different climates.

An advantage of the brushless configuration in which the rotor is inside the stator is that more cross-sectional area is available for the power or 'armature' winding. At the same time the conduction of heat through the frame is improved. Generally an increase in the electric loading is possible (see Section 2.1), providing a greater specific torque. The efficiency is likely to be higher than that of a commutator motor of equal size, and the absence of brush friction helps further in this regard.

The absence of commutator and brushgear reduces the motor length. This is useful not only as a simple space saving, but also as a reduction in the length between bearings, so that for a given stack length the lateral stiffness of the rotor is greater, permitting higher speeds or a longer active length/diameter ratio. This is important in servo-type drives where a high torque/inertia ratio is required. The removal of the commutator reduces the inertia still further.

Commutators are subject to fairly restrictive limits on peripheral speed, voltage between segments, and current density. The maximum speed of the brushless motor is limited by the retention of the magnets against centrifugal force. In small motors with low rotor speeds, the magnets may be bonded to the rotor core, which is usually solid (unlaminated). The bonding must obviously have a wide temperature range and good ageing properties. For high rotor peripheral speeds it is necessary to provide a retaining structure such as a stainless-steel can or a kevlar or wire wrap. This may necessitate an increase in the mechanical airgap, but fortunately the performance is not unduly sensitive to the airgap, which is often twice as large as in induction motors or switched reluctance motors.

The brushless configuration does not come without some disadvantages. The two main disadvantages relative to the commutator motor are (i) the need for shaft position sensing and (ii) increased complexity in the electronic controller. Also, the brushless motor is not necessarily less expensive to manufacture than the commutator motor, which is perhaps slightly more amenable to automated manufacture.

It is important to weigh the advantages and disadvantages of the brushless d.c. motor relative to induction motor drives, which are not only 'brushless' but make use of 'standard' motors. In the same frame, with the same cooling, the brushless PM motor will have better efficiency and power factor, and therefore a greater output power; the difference may be in the order of 20–50 per cent, which is by no means negligible. The power electronic converter required with the brushless motor is similar in topology to the p.w.m. inverters used in induction motor drives. The device ratings may be lower, especially if only a 'constant torque' characteristic is required. Of course, the induction motor can be inexpensively controlled with triacs or

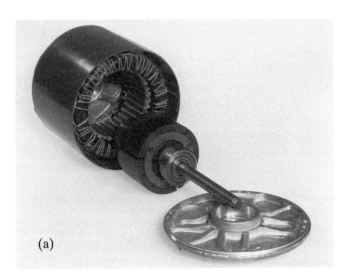

FIG. 4.2. (a) General Electric (USA) three-phase, full-bridge six-pole ECM (brushless d.c.) motor with surface-mounted ferrite magnet arcs on the rotor. Rating is 1/2 hp at 1100 r.p.m.; application is a blower drive used for indoor air handlers and gas furnaces in split-system air-conditioners and heat pumps. Courtesy General Electric Company, Fort Wayne, Indiana, USA.

FIG. 4.2. (b) Brushless d.c. servomotors and controllers. Courtesy Pacific Scientific Motor & Control Division, Rockford, Illinois, USA.

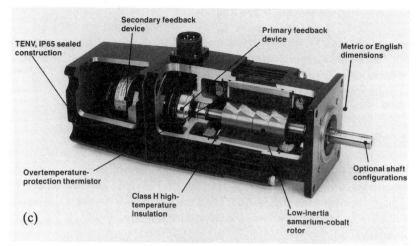

FIG. 4.2. (c) Cutaway of one of the motors shown in Fig. 4.2(b).

FIG. 4.2. (d) Rotor of brushless PM motor with bonded magnet ring for Hall-effect commutation sensor. In other cases the Hall-effect sensor is placed to sense the flux of the main magnets. Courtesy Walter Jones & Co. (Engineers) Ltd., London, UK.

series SCRs, but the performance so obtained is inferior to that of the brushless d.c. system in efficiency, stability, response, and controlled speed range. To obtain comparable performance in the control sense, the induction motor must be fed from a p.w.m. inverter, which is arguably more complex than the brushless PM motor drive. However, the induction motor is capable of operation in the 'field weakening' mode, providing a constant-power capability at high speed. This is difficult to achieve with brushless d.c. motors with surface-mounted rotor magnets.

Something should be said here about the effects of scale. PM excitation is viable only in smaller motors, usually well below 20 kW, and is also subject to certain constraints on the speed range. In very large motors PM excitation does not make sense because the magnet weight (and cost) becomes excessive, while the alternative of electromagnetic excitation either directly (as in the synchronous machine) or by induction (as in the induction motor) becomes relatively more cost-effective.

### 4.2 Magnetic circuit analysis on open-circuit

Figure 4.3(a) shows the cross section of a two-pole brushless d.c. motor having high-energy rare-earth magnets on the rotor. The demagnetization curve of the magnet is shown in Fig. 4.3(b). The axial length of both the stator and the rotor is $l = 50$ mm. First we will consider the open-circuit case, that is with no stator current.

Whenever magnetic circuits are used to analyse a magnetic field, the first task is to identify the main flux paths and assign reluctances or permeances to them. The brushless d.c. motor is very amenable to this kind of analysis. The left half of Fig. 4.3(a) shows the expected flux pattern and Fig. 4.3(c) shows the equivalent magnetic circuit. Only half of the equivalent circuit is shown in Fig. 4.3(c); the lower half is the mirror-image of the upper half about the horizontal axis, which is an equipotential. It is of course permissible to simplify the circuit in this way only if the two halves are balanced. If they are not, the horizontal axis might still be an equipotential but the fluxes and magnetic

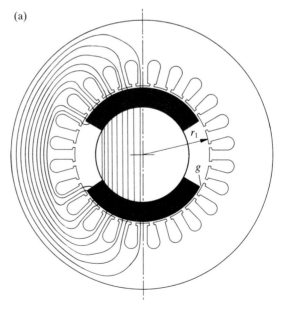

Fig. 4.3. Simple magnetic circuit analysis of BLDC motor. (a) Motor cross section and flux pattern. (b) Magnet demagnetization curve. (c) Magnetic equivalent circuit.

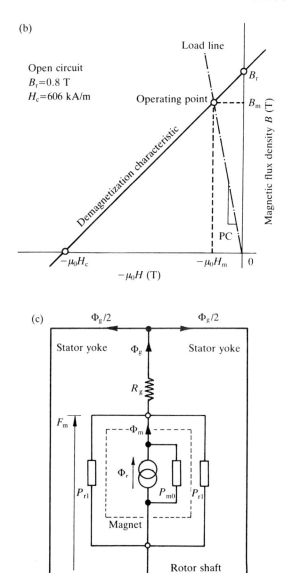

potentials in the two halves would be different, and there could be residual flux in the axial direction, i.e. along the shaft. In practice axial flux is undesirable because it can induce current to flow in the bearings; in some cases this results in accelerated wear.

In the following, the steel core of the stator and the rotor shaft are assumed to be infinitely permeable.

Each magnet is represented by a 'Norton' equivalent circuit consisting of a flux generator in parallel with an internal leakage permeance (Fig. 4.3(c)):

$$\Phi_r = B_r A_m; \quad P_{m0} = \frac{\mu_0 \mu_{rec} A_m}{l_m}$$

where $A_m$ is the pole area of the magnet; $l_m$ is the magnet length in the direction of magnetization (in this case its radial thickness); and $B_r$ is the remanent flux-density. $\mu_{rec}$ is the relative recoil permeability, that is, the slope of the demagnetization curve divided by $\mu_0$. In this case the outer pole area is larger than the inner pole area, but to keep the analysis simple we will take the average. With a magnet arc of 120 degrees,

$$A_m = \frac{2}{3}\pi\left(r_1 - g - \frac{l_m}{2}\right)l.$$

In the example, Fig. 4.3(a), $A_m = 2251.5$ mm$^2$; $\Phi_r = 1.801$ mWb t, and $P_{m0} = 5.942$ E$-7$ Wb/A t.

Most of the magnet flux crosses the airgap via the airgap reluctance $R_g$

$$R_g = \frac{g'}{\mu_0 A_g}$$

where $g'$ is the equivalent airgap length allowing for slotting. The slotting can be taken into account by means of Carter's coefficient, in which case

$$g' = K_c g.$$

Here we will assume $K_c = 1.05$, so that $g' = 1.05$ mm. The airgap area $A_g$ is the area through which the flux passes as it crosses the gap. The precise boundary of this area is uncertain because of fringing, both at the edges of the magnet and at the ends of the rotor. An approximate allowance for fringing can be made by adding $g$ at each of the four boundaries, giving

$$A_g = \left[\frac{2}{3}\pi\left(r_1 - \frac{g}{2}\right) + 2g\right](l + 2g).$$

In the example, $A_g = 2772.3$ mm$^2$, giving $R_g = 3.014$ E5 A t/Wb.

The remaining permeance in the magnetic circuit is the rotor leakage permeance $P_{r1}$, which represents the paths of magnet flux components that fail to cross the airgap. The rotor leakage permeance is difficult to estimate because the flux paths are not obvious. An accurate evaluation of rotor leakage is possible only with numerical techniques such as the finite-element method. With the configuration shown, the rotor leakage permeance is quite small, typically 5–20 per cent of the magnet internal permeance, and it is convenient to include it in a modified magnet internal permeance by writing

$$P_m = P_{m0} + P_{r1} = P_{m0}(1 + p_{r1})$$

where $p_{r1}$ is the normalized rotor leakage permeance, i.e., normalized to $P_{m0}$.

# MAGNETIC CIRCUIT ANALYSIS ON OPEN-CIRCUIT

Thus $p_{r1}$ is of the order 0.05–0.2. In the example, we will assume $p_{r1} = 0.1$.

The magnetic circuit can now be solved. Equating the m.m.f. across the magnet to the m.m.f. across the airgap,

$$F_m = (\Phi_r - \Phi_g)/P_m = \Phi_g R_g, \quad \Phi_g = \frac{\Phi_r}{(1 + P_m R_g)}.$$

If we write the ratio of magnet pole area to airgap area as

$$C_\Phi = \frac{A_m}{A_g}$$

then the airgap flux-density can be extracted as

$$B_g = \frac{C_\Phi}{(1 + P_m R_g)} B_r.$$

In this motor the 'flux concentration factor' $C_\Phi$ is $2251.5/2772.3 = 0.8121$; $P_m R_g = 1.1 \times 5.942\,\text{E}-7 \times 3.014\,\text{E5} = 0.1971$; and hence $B_g = 0.543$ T.

The magnet flux-density $B_m$ can be derived by a similar process:

$$B_m = \frac{1 + P_{r1} R_g}{1 + P_m R_g} B_r$$

giving $B_m = 0.680$ T. Note that the ratio $B_g/B_m = 0.799$, which is a little less than the concentration factor $C_\Phi$, owing to rotor leakage.

Now that we know the magnetic flux-density, it is easy to solve for the magnetizing force $H_m$ in the magnet using the demagnetization characteristic, Fig. 4.3(b). Mathematically the result is

$$-H_m = \frac{B_r - B_m}{\mu_0 \mu_{rec}}\;\text{A/m}$$

and in the example, $H_m = -90.7$ kA/m. The negative sign signifies a demagnetizing force and indicates that the magnet operates in the second quadrant of the $B$–$H$ curve, as expected.

The line drawn from the origin through the operating point in Fig. 4.3(b) is called the 'load-line' and the absolute value of its slope, normalized to $\mu_0$, is called the 'permeance coefficient', PC. The following formula can be derived for PC:

$$\text{PC} = \mu_{rec} \frac{1 + P_{r1} R_g}{P_{m0} R_g}.$$

Alternatively, in terms of geometric dimensions,

$$\text{PC} = \frac{1 + p_{r1} \mu_{rec} C_\Phi \dfrac{g'}{l_m}}{C_\Phi \dfrac{g'}{l_m}}.$$

If Fig. 4.3(b) is drawn with $\mu_0 H_m$ along the x-axis instead of $H_m$, then the units of both axes are the same (T) and the permeance coefficient is the actual slope of the load line. In the example, PC = 5.97. The permeance coefficient is dominated by the ratio of magnet length to effective airgap length, and if $C_\Phi$ is approximately unity and $P_{r1}$ is approximately zero, then it is equal to this ratio. In the example, the ratio is 5/1.05 = 4.76. The permeance coefficient is useful as a measure of how far down the demagnetization curve the magnet operates on open-circuit. It can be shown that

$$\frac{B_m}{B_r} = \frac{PC}{PC + \mu_{rec}}.$$

In the above example, this ratio is 0.85. In motors with a weak flux-concentration factor, the magnet should operate on open-circuit at a high permeance coefficient to maximize the airgap flux-density and the torque per ampere, and to provide adequate margin against demagnetization by armature reaction. Surface-magnet motors generally do have a weak flux concentration factor and a permeance coefficient of six or more is quite normal when using 'hard' magnets (ceramics, rare-earth/cobalt, or neodymium–iron–boron).

The airgap flux-density on open-circuit is plotted in Fig. 4.4. In practice, because of fringing, the distribution is not perfectly rectangular, and there are circumferential as well as radial components of $B$ at the edges of the magnets. Because of the slotting of the stator bore, there will in general be an appreciable ripple superimposed on the calculated waveform. The detailed analysis of all these effects requires a numerical method such as the finite-element method.

The analysis of multiple-pole motors is similar to that of the two-pole motor; using natural equipotentials the magnetic equivalent circuit can be reduced to the per-pole equivalent circuit as was done here.

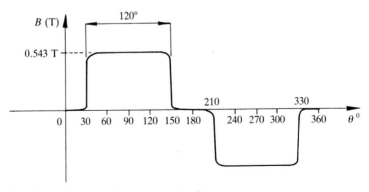

FIG. 4.4. Airgap flux-density on open circuit.

## 4.3 Squarewave brushless motor: torque and e.m.f. equations

The basic torque and e.m.f. equations of the brushless d.c. motor are quite simple, and resemble those of the d.c. commutator motor. The following derivation attempts to encompass several fundamental aspects of these two equations, so as to lay a foundation for understanding the control characteristics and limitations, and the similarities and differences with other machines.

A simple 'concept machine' is shown in Fig. 4.5(a). Note that the two-pole magnet has a pole arc of 180 degrees, instead of the 120 degrees analysed in the previous section. The airgap flux-density waveform is ideally a square wave as shown in Fig. 4.5(b). In practice, fringing causes the corners to be somewhat rounded. The coordinate axes have been chosen so that the centre of a north pole of the magnet is aligned with the x-axis, i.e. at $\theta = 0$.

The stator has 12 slots and a three-phase winding. Thus there are two slots per pole per phase. Each phase winding consists of two adjacent full-pitch coils of $N_1$ turns each, whose axes are displaced from one another by 30 degrees. The winding is a single-layer winding, and any slot contains $N_1$ conductors from only one phase winding. This winding is equivalent, in the active length, to a degenerate concentric winding with only one coil per pole per phase, having a fractional pitch of 5/6. This is a more practical winding than the one analysed because it has less bulky endwindings and is generally easier to assemble. For the same reason, its copper losses are lower.

Consider the flux-linkage $\psi_1$ of coil $a_1A_1$ as the rotor rotates. This is shown in Fig. 4.5(c). Note that $\theta$ now represents the movement of the rotor from the reference position in Fig. 4.5(a). The flux-linkage varies linearly with rotor position because the airgap flux-density set up by the magnet is constant over each pole-pitch of the rotor. Maximum positive flux-linkage occurs at 0 and maximum negative flux-linkage at 180°. By integrating the flux-density around the airgap, the maximum flux-linkage of the coil can be found as

$$\psi_{1\max} = N_1 \int B(\theta) r_1 \, d\theta l = N_1 B_g \pi r_1 l$$

and the variation with $\theta$ as the rotor rotates from 0 to 180° is given by

$$\psi_1(\theta) = \left[1 - \frac{\theta}{\pi/2}\right] \psi_{1\max}.$$

The e.m.f. induced in coil $a_1 A_1$ is given by

$$e_1 = -\frac{d\psi_1}{dt} = -\frac{d\psi_1}{d\theta}\frac{d\theta}{dt} = -\omega \frac{d\psi_1}{d\theta}$$

which gives

$$e_1 = 2N_1 B_g l r_1 \omega \text{ V}.$$

This represents the magnitude of the square-wave e.m.f. $e_{a1}$ shown in

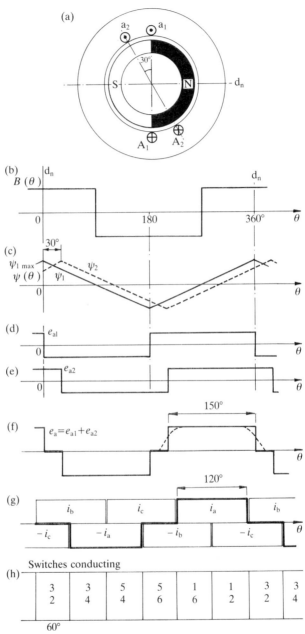

FIG. 4.5. Brushless d.c. motor with ideal waveforms of flux-density, e.m.f., and current. (a) Motor showing two coils of one phase. (b) Magnet flux-density around the airgap. (c) Flux-linkage of coils 1 and 2 as the rotor rotates. (d) e.m.f. waveform of coil 1. (e) e.m.f. waveform of coil 2. (f) e.m.f. waveform of phase a. (g) Ideal phase current waveforms. (h) Switching pattern of switches in the converter of Fig. 4.6(a).

# TORQUE AND E.M.F. EQUATIONS

Fig. 4.5(d). Note that the waveform of e.m.f. in this full-pitch coil with respect to time is an exact replica of the flux-density waveform with respect to position around the rotor in Fig. 4.5(b).

The e.m.f. induced in the second coil of phase A is identical, but retarded in phase by 30°. This is shown in Fig. 4.5(e). If the two coils are connected in series, the total phase voltage is the sum of the two separate coil voltages, and this is shown in Fig. 4.5(f). The basic effect of distributing the winding into two coils is to produce a stepped e.m.f. waveform. In practice, fringing causes its corners to be rounded, as shown by the dotted lines. The waveform then has the 'trapezoidal' shape that is characteristic of the brushless d.c. motor. With 180° magnet arcs and two slots per pole per phase, the flat top of this waveform is ideally 150° wide, but in practice the fringing field reduces this to a somewhat smaller value, possibly as little as 120°. The magnitude of the flat-topped phase e.m.f. is given by

$$e = 2N_{ph}B_g l r_1 \omega \text{ V}$$

where $N_{ph}$ is the number of turns in series per phase. In this case

$$N_{ph} = 2N_1$$

because the two coils considered are assumed to be in series. In a machine with $p$ pole-pairs, the equation remains valid provided $N_{ph}$ is the number of turns in series per phase and $\omega$ is in mechanical radians per second.

Figure 4.5(g) shows an ideal rectangular waveform of phase current, in which the current pulses are 120 electrical degrees wide and of magnitude $I$. The positive direction of current is against the e.m.f., that is, positive current is motoring current. The conduction periods of the three phases are symmetrically phased so as to produce a three-phase set of balanced 120° square waves. If the phase windings are star-connected, as in Fig. 4.6(a), then at any time there are just two phases and two transistors conducting.

During any 120° interval of phase current the instantaneous power being converted from electrical to mechanical is

$$P = \omega T_e = 2eI.$$

The '2' in this equation arises from the fact that two phases are conducting. Using the expression derived above for the e.m.f., the electromagnetic torque is given by

$$T_e = 4N_{ph}B_g l r_1 I \text{ N m.}$$

This equation is valid for any number of pole-pairs. The similarity between the brushless motor and the commutator motor can now be seen. Writing $E = 2e$ to represent the combined e.m.f. of two phases in series, the e.m.f. and torque equations can be written in the form

$$E = k\Phi\omega \text{ and } T = k\Phi I$$

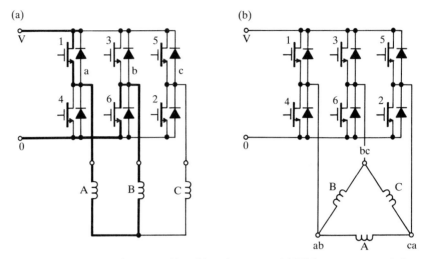

FIG. 4.6. Converter or inverter of brushless d.c. motor. (a) With star-connected phase windings; (b) with delta-connected phase windings.

where

$$k = \frac{4}{\pi} N_{ph} \text{ and } \Phi = B_g r_1 \pi l.$$

$k$ is the 'armature constant' and $\Phi$ is the flux. These equations for e.m.f. and torque are exactly the same as for the d.c. commutator motor; only the form of the constant $k$ is different. It is clear that with ideal waveshapes and with perfect commutation, these equations are true at all instants of time. The electronic commutation of the converter switches has thus assumed the function of the mechanical commutator in the commutator motor, to give a pure 'd.c.' machine with constant, ripple-free torque.

In practice, of course, none of the ideal conditions can be perfectly realized. The main result of this is to introduce ripple torque, but the basic relationships of e.m.f. proportional to speed and torque proportional to current remain unchanged.

## 4.4 Torque/speed characteristic: performance and efficiency

The torque/speed curve of the ideal brushless motor can be derived from the foregoing equations. If the commutation is perfect and the current waveforms are exactly as shown in Fig. 4.5(g), and if the converter is supplied from an ideal direct voltage source $V$, then at any instant the following equation can be written for the d.c. terminal voltage:

$$V = E + RI$$

where $R$ is the sum of two phase resistances in series and $E$ is the sum of two phase e.m.f.s in series. This equation is exactly the same as that of the commutator motor. The voltage drops across two converter switches in series are omitted, but they correspond exactly to the two brush voltage drops in series in the commutator motor.

Using this equation together with the e.m.f. and torque equations, the torque/speed characteristic can be derived as:

$$\omega = \omega_0 \left[1 - \frac{T}{T_0}\right]$$

where the no-load speed is

$$\omega_0 = \frac{V}{k\Phi} \text{ rad/sec}$$

and the stall torque is given by

$$T_0 = k\Phi I_0.$$

This is the torque with the motor stalled, i.e. at zero speed. The stall current is given by

$$I_0 = \frac{V}{R}.$$

This characteristic is plotted in Fig. 4.7. If the phase resistance is small, as it should be in an efficient design, then the characteristic is similar to that of a shunt d.c. motor. The speed is essentially controlled by the voltage $V$, and may be varied by varying the supply voltage. The motor then draws just enough current to drive the torque at this speed. As the load torque is increased, the speed drops, and the drop is directly proportional to the phase resistance and the torque.

The voltage is usually controlled by chopping or p.w.m. (Section 4.8). This gives rise to a family of torque/speed characteristics as shown in Fig. 4.7. Note the boundaries of continuous and intermittent operation. The continuous limit is usually determined by heat transfer and temperature rise. The intermittent limit may be determined by the maximum ratings of semiconductor devices in the controller, or by temperature rise.

In practice the torque/speed characteristic deviates from the ideal form because of the effects of inductance and other parasitic influences. Detailed consideration of these requires computer simulation (see Section 4.9). Figure 4.7 also shows the possibility of extending the speed range by

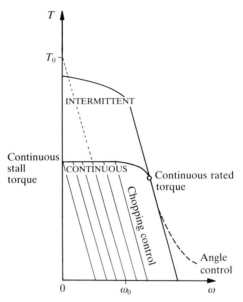

FIG. 4.7. Torque/speed characteristic of ideal brushless d.c. motor.

advancing the phase and/or the dwell of the conduction period relative to the rotor position [see Jahns 1984].

## 4.5 Alternative formulations for torque and e.m.f.

Most brushless d.c. motors have PM-excited rotors, but there is no theoretical reason why the rotor could not be a wound rotor. The rotor current must be d.c. It can be supplied through slip-rings or, in the 'brushless' configuration, from a rotating rectifier which is fed with a.c. from an exciter on the same shaft. The exciter has a rotating a.c. winding and a stationary d.c. excitation winding or permanent magnet. This configuration is that of the brushless synchronous machine, which can be regarded as the big brother of the brushless PM motor; it is usually found in larger ratings as a sinewave motor (see Chapter 5).

The point here is to show the theoretical relationship between a wound rotor and a PM rotor. If the rotor is wound, and if both stator and rotor are cylindrical, (i.e. no salient poles) then the instantaneous electromagnetic torque produced with one stator phase excited is given by

$$T = i_1 i_2 \frac{dL_{12}}{d\theta}$$

where '$_1$' refers to the stator winding and '$_2$' refers to the rotor winding. A permanent magnet behaves like a winding with fixed d.c. current, provided that its demagnetization curve and load line (airgap line) remain constant. The load line will remain constant as long as the stator is cylindrical and there are no perturbing effects such as rotor eccentricity. The demagnetization curve of the magnet will remain constant provided the temperature remains constant. With these conditions met, the flux-linkage $\psi_M$ in the stator winding produced by the magnet can be written in terms of the current $i_2$ in a fictitious rotor winding, which is assumed to have a mutual inductance $L_{12}$ with the stator winding: thus

$$\psi_M = L_{12} i_2.$$

With fixed rotor current, the torque can be written

$$T = i_1 \frac{d}{d\theta}(L_{12} i_2) = i_1 \frac{d\psi_M}{d\theta}.$$

This equation is helpful when comparing the operation of PM machines with reluctance machines, as we shall see later. It is only true when the magnet flux is fixed and the flux-linkage varies as a function of rotor angle. Otherwise the stored field energy changes and other torque components appear. This condition will be satisfied if the flux is established by the magnet and there is no 'perturbing' field due to armature reaction. In motors with 'hard' surface-mounted rotor magnets the permeability of the magnets to armature-reaction flux is the recoil permeability, whose relative value is nearly unity, so that the magnet effectively presents a large airgap to the armature m.m.f. Under these conditions the main flux is liable to be affected by armature m.m.f. to only a small degree. The magnet length is usually made large enough to satisfy this condition, partly to maximize the torque per ampere, but also to avoid local demagnetization of the magnet.

We have seen in Section 4.3 that for a single full-pitch coil,

$$\psi_M = \left[1 - \frac{2\theta}{\pi}\right] N_1 B_g r_1 \pi l, \qquad 0 < \theta < \pi$$

and hence

$$T = -2 N_1 B_g r_1 l i_1.$$

If the current has a 120° square waveform of amplitude $I$, this expression leads to the same one derived in Section 4.3. The negative sign arises because positive current corresponds to generator action, but it is more convenient to use its negative, which is positive for motoring, as in Section 4.3 and Fig. 4.5; this removes the negative sign in the torque expression.

The product $N_1 i_1$ is the total current in the slot, and this formula can be interpreted as

$$T = 2 f r_1$$

where

$$f = B_g(N_1 I)l$$

and is the force acting on each slot current in the circumferential direction. (The use of this 'BIL' formulation for force and torque is also permissible in the commutator motor.) It shows that the greatest torque per ampere will be obtained if the current is caused to flow in conductors located in the region of maximum flux density; that is, towards the centre of the magnet poles. The best current pattern is thus in 'space quadrature' with the magnetic field of the magnets. If these currents were acting alone, they would produce a magnetic field with poles on the stator bore, displaced exactly 90 electrical degrees from those of the magnets. In squarewave brushless motors this space quadrature cannot be realized continuously because of the stepped rotation of the armature m.m.f. (see Section 4.6); and because of time delays caused by inductance.

The e.m.f. equation can be derived in a similar way. From Faraday's law,

$$e = -\frac{d\psi_M}{dt} = -\frac{d\psi_M}{d\theta}\omega$$

$$= \frac{2}{\pi} N_1 B_g r_1 \pi l \omega$$

$$= 2 N_1 B_g l v.$$

This is equivalent to the sum of e.m.f.s $Blv$ in $2N_1$ conductors in series, where $v$ is their velocity through the magnetic field. $B$ is established entirely by the magnets, and includes no component of armature-reaction field.

## 4.6 Motors with 120° and 180° magnet arcs: commutation

In Section 4.3 we developed the concept of the ideal brushless machine with 180° magnet arcs and 120° square-wave currents. The 180° magnet arcs were assumed to produce a rectangular distribution of flux density in the airgap, as in Fig. 4.5(b). The phase windings were assumed to be star-connected, as in Fig. 4.6(a).

This configuration is represented in a different way in Fig. 4.8(a). The rotor magnet poles are shaded to distinguish north and south. The phase belts are shaded as complete 60° sectors of the stator bore. In the ideal motor considered in Section 4.3, there are two slots in each of these phase belts (see Fig. 4.5). The currents in these two slots are identical, and the conductors in them are normally in series. Between the rotor ring and the stationary 'phasebelt ring' in Fig. 4.8(a) is a third ring called the 'm.m.f. ring'. This represents the m.m.f. distribution of the stator currents at a particular instant.

MOTORS WITH 120° AND 180° MAGNET ARCS: COMMUTATION    71

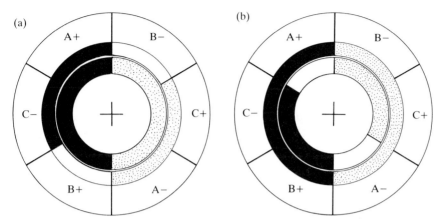

FIG. 4.8. (a) BLDC motor with 180° magnet arcs and 120° square-wave phase currents.

FIG. 4.8. (b) BLDC motor with 120° magnet arcs and 180° square-wave phase currents.

With a star winding and two phases conducting, the m.m.f. distribution is such that at any instant there are two sectors of opposite polarity, each 120 electrical degrees wide, separated by two 60° sectors with zero m.m.f. At the instant shown, phase A is conducting positive current and phase C is conducting negative current. The resulting m.m.f. distribution has the same shading as the N and S rotor poles to indicate the generation of torque. Where the m.m.f. distribution and the rotor flux-density distribution have like shading, positive torque is produced (assumed clockwise). Where the m.m.f. and flux shadings are unlike, negative torque is produced. Where either one is zero, no torque is produced. The total torque is, of course, the 'integral' of the contributions from around the entire airgap periphery.

It is very helpful to construct a model of Figs 4.8(a) and (b) using construction paper, in which the three separate rings, suitably shaded or coloured, are cut out and pinned at the centre so that the two inner ones can be independently rotated while the outer 'phasebelt' ring remains stationary. This model will prove invaluable in sorting out the sequence of events as the rotor rotates.

In the position shown in Fig. 4.8(a), positive torque is being contributed by phases A and C. Only $\frac{2}{3}$ of the magnet arc and $\frac{2}{3}$ of the stator conductors are contributing torque at this instant. As the rotor moves forward in the clockwise direction, with fixed currents in the phasebelts, the overlap between the m.m.f. and flux distributions remains fixed at 120 electrical degrees. Both the flux and m.m.f. distributions are assumed to be rectangular with fixed magnitudes, and therefore the torque remains constant for 60° of rotation. At this point the trailing edges of the rotor poles start to 'uncover' the C phasebelts, and the torque contribution of the phase C starts to decrease linearly. However, during this 60° rotation the leading edges of the magnet poles have been 'covering' the B phasebelts with flux of the correct polarity, such that if the (negative) current is commutated from phase C to phase B exactly at the 60° point, then the torque will be unaffected and will continue constant for a further 60°. At that point the rotor will be advanced 120° from the position shown in Fig. 4.8(a), and the positive current must be commutated from phase A to phase C. The rotating model makes these transitions easier to visualize.

The phase current waveforms for positive rotation are shown in Fig. 4.9, together with a series of 'vector' diagrams depicting the m.m.f. formation in each 60° period. The phase currents and the line currents are 120° square waves. The switching of the switches in the converter is summarized in Table 4.1(a). Note that each source and each sink device in each phaseleg is on for 120° and off for 240°. (In a multiple-pole motor, these measures are, of course, electrical degrees).

The production of smooth, ripple-free torque depends on the fact that the magnet pole arc exceeds the m.m.f. arc by 60°. The magnet is therefore able to rotate 60° with no change in the flux-density under either of the conducting

**Table 4.1.** Commutation tables for three-phase brushless d.c. motors
(a) 180° Magnet-Star winding
    120° square-wave phase currents

| Rotor position | A | B | C | aU (1) | aL (4) | bU (3) | bL (6) | cU (5) | cL (2) |
|---|---|---|---|---|---|---|---|---|---|
| 0–60    | +1 | 0  | −1 | 1 | 0 | 0 | 0 | 0 | 1 |
| 60–120  | +1 | −1 | 0  | 1 | 0 | 0 | 1 | 0 | 0 |
| 120–180 | 0  | −1 | +1 | 0 | 0 | 0 | 1 | 1 | 0 |
| 180–240 | −1 | 0  | +1 | 0 | 1 | 0 | 0 | 1 | 0 |
| 240–300 | −1 | +1 | 0  | 0 | 1 | 1 | 0 | 0 | 0 |
| 300–360 | 0  | +1 | −1 | 0 | 0 | 1 | 0 | 0 | 1 |

(b) 120° Magnet-Delta winding
    180° square-wave phase currents

| Rotor position | A | B | C | abU (1) | abL (4) | bcU (3) | bcL (6) | caU (5) | caL (2) |
|---|---|---|---|---|---|---|---|---|---|
| 0–60    | +1 | +1 | −1 | 0 | 0 | 1 | 0 | 0 | 1 |
| 60–120  | +1 | −1 | −1 | 1 | 0 | 0 | 0 | 0 | 1 |
| 120–180 | +1 | −1 | +1 | 1 | 0 | 0 | 1 | 0 | 0 |
| 180–240 | −1 | −1 | +1 | 0 | 0 | 0 | 1 | 1 | 0 |
| 240–300 | −1 | +1 | +1 | 0 | 1 | 0 | 0 | 1 | 0 |
| 300–360 | −1 | +1 | −1 | 0 | 1 | 1 | 0 | 0 | 0 |

Numbers in brackets refer to transistors in Figs 4.6(a) and (b).

# MOTORS WITH 120° AND 180° MAGNET ARCS: COMMUTATION 73

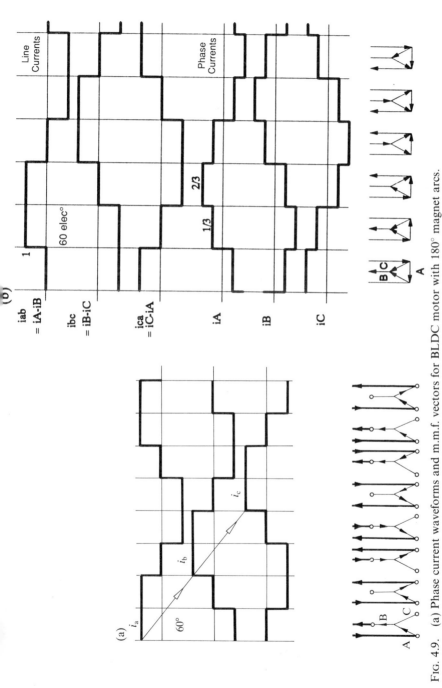

FIG. 4.9. (a) Phase current waveforms and m.m.f. vectors for BLDC motor with 180° magnet arcs.
FIG. 4.9. (b) Line and phase current waveforms and m.m.f. vectors for BLDC motor with 120° magnet arcs.

phasebelts. An inevitable result of this is that only $\frac{2}{3}$ of the magnet and $\frac{2}{3}$ of the stator conductors are active at any instant.

In a practical motor the magnet flux-density distribution cannot be perfectly rectangular, as shown in Fig. 4.5(b). Even with highly coercive magnets and full 180° magnet arcs, there is a transition section of the order of 10–20° in width. This 'fringing' effect is similar to that described in connection with Fig. 4.4. Likewise on the stator side, the m.m.f. distribution is not rectangular but has a stepped waveform that reflects the slotting. To some extent these two effects cancel each other, so that satisfactory results are achieved with a magnet arc as short as 150° and two slots per pole per phase (i.e., two slots per phasebelt). But there is always a dip in the torque in the neighbourhood of the commutation angles. This torque dip occurs every 60 electrical degrees, giving rise to a torque ripple component with a fundamental frequency equal to $6p$ times the rotation frequency, where $p$ is the number of pole-pairs. The magnitude and width of the torque dip may be exaggerated further by the time it takes to commutate the phase current from one phaseleg to another; this transition depends on the phase inductance and the available voltage.

It is important to note, in relation to the commutation of the converter, that the back-e.m.f. is essentially at its full value when the outgoing phase is commutated, and that this e.m.f. is in such a direction as to drive the current down and assist commutation. On the other hand, the back-e.m.f. is of equal value in the incoming phase and is in such a direction as to oppose the current build-up. This is the primary reason for the characteristic asymmetry in the phase current waveform, an example of which is shown in Fig. 4.10(a). (This waveform corresponds to full-voltage operation with no p.w.m.)

While the flux-distribution of the magnet rotates with the rotor in a continuous fashion, the m.m.f. distribution of the stator remains stationary for 60° and then jumps to a position 60° ahead. This motor is not a rotating-field machine in the sense associated with a.c. machines.

An alternative form of the ideal brushless motor is shown in Fig. 4.8(b), and

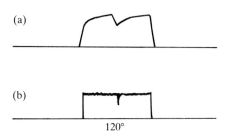

FIG. 4.10. Phase current waveforms. (a) High speed, full voltage. Note the dip caused by commutation of the other two phases. (b) Low speed, with current controlled by chopping.

## MOTORS WITH 120° AND 180° MAGNET ARCS: COMMUTATION

again a rotating cut-out model is helpful to visualize the state transitions as the rotor rotates. In this case the magnet arc is only 120° (electrical). In order to produce a smooth, ripple-free torque using the same 'constant overlap' principle as before, the stator m.m.f. distribution must now be a rectangular square wave having 180° of positive m.m.f. and 180° of negative m.m.f. at the airgap. From the starting position shown in Fig. 4.8(b), if the m.m.f. distribution remains fixed then the rotor can rotate 60° clockwise before any change is necessary in the m.m.f. distribution. At that point the m.m.f. distribution must be switched forwards by 60° and the process continues. As before, with ideal flux and m.m.f. distributions and perfect commutation, the torque is constant.

It is clear from Fig. 4.8(b) that all the stator conductors are excited with current at any instant, but that only $\frac{2}{3}$ of them are producing torque—those that are 'covered' by a rotor pole. Compared with the arrangement of Fig. 4.8(a), if the ampere-conductors per slot and the peak flux-density are kept the same, the motor of Fig. 4.8(b) has 1.5 times the copper losses, but produces the same torque with only $\frac{2}{3}$ the magnet material. Actually the ampere-conductors per slot would have to be reduced in Fig. 4.8(b), because the duty cycle is 1.0 instead of $\frac{2}{3}$. Therefore the motor of Fig. 4.8(b) is likely to be less efficient than that of Fig. 4.8(a). Offsetting this disadvantage is the fact that for the same magnet flux-density, the total flux in Fig. 4.8(b) is only $\frac{2}{3}$ that in Fig. 4.8(a), so that only $\frac{2}{3}$ of the stator yoke thickness is required. If the stator outside diameter is kept the same, the slots can be made deeper so that the loss of ampere-conductors can be at least partially recovered. Consequently the efficiency of the motor of Fig. 4.8(b) may not be very much lower than that of Fig. 4.8(a).

The stator m.m.f. distribution of Fig. 4.8(b) cannot be realized with a star-connected winding. Instead the windings must be connected in delta. The phase and line current waveforms are shown in Fig. 4.9(b), together with a set of m.m.f. 'vector' diagrams. Each phase current is a 180° square wave. However, each line current is a 120° square wave, as in the case of Figs 4.8(a) and 4.9(a). The corresponding commutation table is given in Table 4.1(b), and the converter connections in Figs 4.6(a) and (b). Because the line currents are 120° square waves in both cases, the same converter and indeed the same commutation logic can be used for either motor configuration. Integrated-circuit drivers for brushless d.c. motors generally do not distinguish between star- and delta-connected motors and are capable of driving both, using the same logic to decode the shaft position sensor signals (usually derived from Hall-effect sensors).

The operation of the delta-connected motor merits a little closer attention. Consider the period 0–60° in Table 4.1(b). Referring to Fig. 4.8(b), only the C phasebelts remain 'covered' by the magnet poles during this period. The coverage of each A phasebelt is increasing linearly, while that of each B phasebelt is decreasing at the same rate. Since all the conductors are carrying the same current (with appropriate polarity), the increasing torque contribution of phase A is balanced by the decreasing contribution of phase B, and the total torque remains constant. Similarly, during this period there is a linear increase in the back-e.m.f. in phase A, and an equal and opposite decrease in the back-e.m.f. in phase B, such that the total e.m.f. of phases A and B in series between terminals bc and ca is $e$, that is, the same as the (constant) e.m.f. of phase C. It is for this reason that the line current divides equally between the two parallel paths, even though one path contains only one phase winding (C) while the other contains two in series (A and B). In practice, of course, this balance is slightly imperfect because of the resistances and inductances of the windings. But the preservation of balanced currents in the two parallel paths throughout each 60° period is important, because circulating currents would produce excessive torque ripple and additional losses.

Just as in the motor of Fig. 4.8(a), the effects of fringing flux, slotting, and commutation overlap combine to produce torque ripple.

In this discussion we have used the NBLV and NBIL concepts for e.m.f. and torque respectively, but the more rigorous concepts of changing flux-linkage and energy balance could equally well be used to analyse the operation, and indeed would be more convenient for the purposes of circuit analysis and simulation. The relationships between these formulations are shown in Section 4.5. In terms of the 'coverage' of a phasebelt by magnet pole flux, the e.m.f. of a winding is changing linearly whenever the coverage of its phasebelts is changing; that is, when a leading or trailing edge of a rotor pole is sweeping across them. The change in e.m.f. can be ascribed to the increasing number of conductors coming under the influence of the magnet pole flux. During these 60° intervals the flux-linkage of the winding is changing quadratically. If the coverage is fixed, as it is in Fig. 4.8(b) for the C phasebelts for the imminent 60° of rotation, then the e.m.f. is constant and there is a linear change in the flux-linkage. These considerations apply only to the idealized motors of Fig. 4.8. In practice the form of the flux-linkage, e.m.f., current, and torque waveforms is more complicated and can only be calculated by computer simulation.

## 4.7 Squarewave motor: winding inductances and armature reaction

Figure 4.11 shows the magnetic flux established by a full-pitch winding with one slot per pole per phase. The total m.m.f. around a complete loop or flux-line is equal to $Ni$, where $N$ is the number of conductors in the slot and $i$ is the current. If the steel in the rotor and stator is assumed to be infinitely permeable, then the m.m.f. is concentrated entirely across the two airgaps.

# WINDING INDUCTANCES AND ARMATURE REACTION

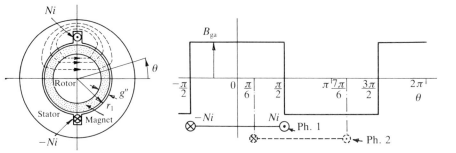

FIG. 4.11. Winding inductance of squarewave motor with one slot per pole per phase.

Across each airgap the m.m.f. drop is $Ni/2$. If the flux is assumed to be radial in the gap, the magnetizing force in each gap is

$$H = \frac{NI}{2g''}.$$

In a surface-magnet motor the gap $g''$ includes the radial thickness of the magnet as well as the physical airgap, which may be modified by Carter's coefficient for the stator slotting (as in Section 4.2). A reasonable approximation for $g''$ is

$$g'' = g' + \frac{l_m}{\mu_{rec}}$$

where

$$g' = K_c g.$$

The flux-density produced by this magnetizing force at the bore of the stator is

$$B_{ga} = \frac{\mu_0 Ni}{2g''}.$$

The subscript 'a' is added to emphasize that the flux is that which is excited by armature reaction, i.e., by the stator current. The flux distribution around the airgap is plotted in Fig. 4.11. The flux-linkage of the coil is

$$\psi = Nlr_1 \pi B_{ga}$$

and the self-inductance is

$$L_g = \frac{\psi}{i} = \frac{\pi \mu_0 N^2 l r_1}{2g''} \text{ H}.$$

The subscript '$g$' has been added to emphasize that this inductance is associated with flux in the airgap. It is necessary to add the slot-leakage flux and the end-turn leakage flux, but these will be treated later. This result has been derived for a two-pole machine. If there are $p$ pole-pairs and if all the turns are in series, then the number of turns in a phase winding is

$$N_{ph} = Np$$

and

$$L_g = \frac{\pi \mu_0 N_{ph}^2 l r_1}{2p^2 g''} \text{ H.}$$

For example, in the two-pole machine of Fig. 4.3(a), with 60 turns per phase, a stack length of 50 mm, a stator bore of 50 mm, a magnet thickness of 5 mm, and an effective airgap of 1.05 mm, if the relative recoil permeability is 1.05, then

$$g'' = 1.05 + \frac{5}{1.05} = 5.81 \text{ mm}$$

and

$$L_g = \frac{\pi(4\pi \times 10^{-7}) \times 60^2 \times 50 \times 25 \times 10^{-6}}{2 \times 5.81 \times 10^{-3}} = 1.529 \text{ mH.}$$

The mutual inductance between two phases whose axes are displaced by 120 electrical degrees is important. This can be calculated by adding up the flux-linkage of a second coil placed in the field of the first one. For a second coil with conductors located at the angles $90° + 120° = 210°$ and $-90° + 120° = 30°$, the flux-linkage is

$$\psi_2 = Nlr_1 B_g \left[ -\left(\frac{7\pi}{6} - \frac{\pi}{2}\right) + \left(\frac{\pi}{2} - \frac{\pi}{6}\right) \right].$$

Note how the positive and negative signs account for the direction of the flux-density in the airgap. Substituting the expression for $B_g$ from above, we get

$$M_g = \frac{\psi_2}{i} = -\frac{1}{3} \frac{\pi \mu_0 N^2 l r_1}{2g''} \text{ H.}$$

Note that

$$\frac{M_g}{L_g} = -\frac{1}{3}.$$

Once $L_g$ has been calculated, this equation can be used for $M_g$. Physically this result makes sense: if the second coil was aligned with the first, the self and mutual would be equal. If the second coil was displaced 90° from the first, the mutual would be zero. With a rectangular flux distribution produced by current in the first coil, and with concentrated windings of one slot per pole per phase, the flux-linkage of the second one varies linearly with the angle between their axes. By rotating the second coil to a position one-third of a right angle past the 90° position, it picks up or links $\frac{1}{3}$ of the flux, but in the negative direction relative to its own positive axis. If there are $p$ pole-pairs with all turns

# WINDING INDUCTANCES AND ARMATURE REACTION

in series, then $N$ is substituted by $N_{ph}/p$. If there are '$a$' parallel paths through the winding, then both the self and mutual inductances are multiplied by $1/a^2$.

Now consider a winding with two slots per pole per phase, as shown in Fig. 4.12. For example, in a four-pole, three-phase machine there would be 24 slots with a slot-pitch of 15°. The conductors of a second, unexcited phase are shown in their correct positions for calculating the mutual inductance between phases. By a process similar to the one above, in which the flux-linkages of the individual coils are added together, the self inductance is shown to be

$$L_g = \frac{\pi \mu_0 N_{ph}^2 l r_1}{2 p^2 g''} k_w$$

where

$$k_w = 1 - \frac{1}{3q}$$

is the winding factor for $q$=two slots per pole per phase, i.e. 0.833. The self-inductance is thus only 83.3 per cent of the value which would be obtained with the same number of turns per phase concentrated in one slot per pole per phase.

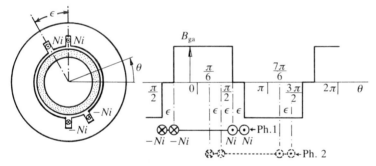

FIG. 4.12. Winding inductance of squarewave motor with two slots per pole per phase.

When the mutual inductance is evaluated, using the same method as before, it is found that the distribution of the second winding cancels the effect of the step in the flux distribution, so that the actual value of the mutual inductance is the same as with one slot per pole per phase (provided the total turns are the same). The ratio between the self and mutual inductances is therefore

$$\frac{M_g}{L_g} = \frac{-1/3}{k_w} = -0.4.$$

Once $L_g$ has been calculated, this equation can be used for $M_g$.

Now consider a winding with three slots per pole per phase, as shown in Fig. 4.13. For example, in a four-pole three-phase machine there would be 36

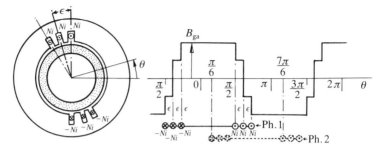

FIG. 4.13. Winding inductance of squarewave motor with three slots per pole per phase.

slots with a slot-pitch of 10°. The conductors of a second, unexcited phase are shown in their correct positions for calculating the mutual inductance between phases. By a process similar to the one above, in which the flux-linkages of the individual coils are added together, the self inductance is shown to be

$$L_g = \frac{\pi \mu_0 N_{ph}^2 l r_1}{2 p^2 g''} k_w$$

where

$$k_w = 1 - \frac{16}{27q}$$

is the winding factor for $q=$ three slots per pole per phase, i.e. 0.802. The self-inductance is thus only 80.2 per cent of the value which would be obtained with the same number of turns per phase concentrated in one slot per pole per phase.

When the mutual inductance is evaluated, using the same method as before, it is again found that the distribution of the second winding cancels the effect of the steps in the flux distribution, so that the actual value of the mutual inductance is the same as with one slot per pole per phase (provided the total turns are the same). The ratio between the self and mutual inductances is therefore

$$\frac{M_g}{L_g} = \frac{-1/3}{k_w} = -0.415.$$

Once $L_g$ has been calculated, this equation can be used for $M_g$.

## 4.8 Controllers

The general structure of a controller for a brushless PM motor is shown in Fig. 4.14. This schematic diagram shows the functions required to control the

# CONTROLLERS

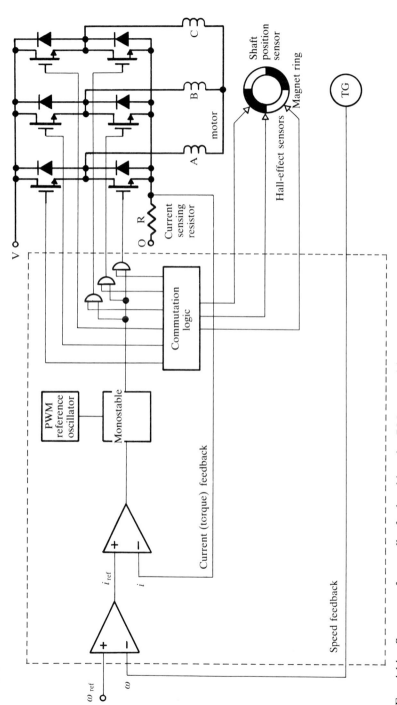

FIG. 4.14. Structure of controller for brushless d.c. PM motor drive.

drive in the 'chopping control' range of Fig. 4.7, i.e., the commutation angles are fixed.

The rotor shaft position is sensed by a Hall-effect sensor, a slotted optical disk, or some other transducer, providing signals as represented in Table 4.1. These signals are 'decoded' by combinatorial logic to provide the firing signals for 120° conduction on each of the three phases. The commutation logic or 'rotor position decoder' therefore has six outputs which control the upper and lower phaseleg transistors. Programmable logic arrays, gate arrays, and EPROMS are all suitable for this function. In general there will be level-shifting circuits to buffer the outputs of the logic circuit and provide the drive to the power devices.

The basic forward control loop is a voltage control, implemented in Fig. 4.14 by a monostable clocked at a fixed reference frequency, which is typically a few kHz. In general it is desirable to make this frequency ultrasonic to minimize noise. The duty-cycle or off-time is controlled by an analogue voltage reference that represents the desired speed. In Fig. 4.14 the p.w.m. is applied only to the lower phaseleg transistors. This not only reduces the current ripple but also avoids the need for wide bandwidth in the level-shifting circuit that feeds the upper phaseleg transistors. With higher d.c. supply voltages this can be a useful saving. The upper transistors need not be of the same type as the lower ones and need only switch at the commutation frequency, i.e. only a few hundred Hz. Note the use of AND gates as a simple way of combining the commutation and chopping signals to the lower transistors. (In practice these would usually be implemented in negative TTL or CMOS logic).

From a control point of view the brushless motor is similar to the d.c. commutator motor, as the simple torque and voltage equations show. Consequently it is possible to implement current (torque) feedback and speed feedback in the same way as for the d.c. motor, and generally this results in a well-behaved system although compensation may be necessary in either or both loops to improve stability and transient response. A tight speed control is thus possible over a wide range of speed and torque using relatively simple techniques that are familiar with commutator motors. This is in contrast, to some degree, with the a.c. induction motor, which cannot accommodate speed and torque control loops in this form without having reference-frame transformations, as in the field-oriented control technique, or their equivalent. The same is true of the hybrid PM and synchronous reluctance motors in Chapter 6. The switched reluctance motor is capable of being controlled by a circuit similar to Fig. 4.14, but a problem arises because of the strongly inverse torque/speed characteristic, and current-mode control or compensation of the torque loop may be necessary.

Sometimes the instantaneous current in the brushless PM motor is regulated in each phase by a hysteresis-type regulator which maintains the current within adjustable limits. This is called 'current-mode' control and several algorithms are possible to control the switching. In this case current sensors are needed in each phase, and their bandwidth must obviously be considerably wider than that of the sensing resistor shown in Fig. 4.14. The speed feedback signal, derived in Fig. 4.14 from a tachometer-generator TG, can also be derived from the shaft position sensor by a frequency-to-voltage converter. This technique only works at high speeds.

Many of the functions of the circuit in Fig. 4.14 can be performed digitally, and it is increasingly common to have a serial communications interface that permits the system to be computer controlled. In high-performance systems the shaft position sensor may be a resolver or optical encoder, with special-purpose decoding circuitry. At this level of control sophistication, it is possible to fine-tune the firing angles and the p.w.m. control as a function of speed and load, to improve various aspects of performance such as efficiency, dynamic performance, or speed range.

There is a wide range of integrated circuits available with many of the functions in Fig. 4.14. The functions inside the dotted line are all found on many such products, as well as many protective functions such as over- and under-voltage protection, overcurrent protection, and lockout protection (preventing the upper and lower switches in one phaseleg from turning on at the same time and causing a 'shoot-through' failure). Some 'smartpower' products also include driver stages with outputs of 5 A at 50 V, or more.

## 4.9 Computer simulation

As part of the design process it is often necessary to simulate the performance of a brushless d.c. drive on the computer. At the simplest level, it can be modelled as a d.c. motor with its speed and torque control loops, and we have seen that the form of the equations is identical.

Computer simulation can also be used as part of the design process for the motor and controller together, (see Section 7.5 and Fig. 7.24). The voltage equations and the equations of motion are ordinary differential equations that can be integrated by standard techniques such as Runge-Kutta. While this procedure has been widely used for analysis of drives that have been already designed, it is now possible to incorporate such simulations into the original design process, even on the PC, and in this way the combined design of the motor and controller can be done at the same time.

The integration of the voltage equations must take account of the circuit topology, which changes at every commutation. With reference to Figs 4.5 and 4.6, consider a 60-degree period when current is flowing through phases A and B of a star-connected motor. There is one main voltage equation in this 'two-phases-on' connection, corresponding to the mesh which includes the supply, transistor 1, phase windings A and B, transistor 6, and back to the supply. Writing the mesh current as

$$i_1 = i_a = -i_b,$$

the voltage equation is

$$V = 2V_Q + 2Ri_1 + 2L\frac{di_1}{dt} + e_a - e_b.$$

Here $e_a$ and $e_b$ are the instantaneous values of the open-circuit voltages on phases A and B. These can be calculated from a magnetostatic analysis of the flux-linkage of the phase windings as a function of the rotor angle $\theta$, which is known from integrating the equations of motion:

$$\frac{d\omega}{dt} = \frac{1}{J}(T_m - T_L)$$

$$\frac{d\theta}{dt} = \omega.$$

In the simplest case the speed is assumed constant, and no integration is necessary for the mechanical equations. The rotor position is determined from the relationship

$$\theta = \omega t + \theta_0$$

where $\theta_0$ is the initial value at the start of the integration. The main loop voltage equation can be rearranged and integrated along with the mechanical equations:

$$\frac{di_1}{dt} = \frac{1}{2L}[V - 2V_Q - 2Ri_1 - (e_a - e_b)]$$

where $V_Q$ is the voltage drop across each transistor. If the transistors are field-effect devices, it may be more appropriate to model them in terms of their on-state resistance, which can be included in the resistance $R$. This formulation provides for mutual inductance effects if they are included in the value of the inductance term $2L$. If the motor is delta-connected, it may be necessary to set up a second mesh voltage equation for the delta, and mutual coupling terms must be included in both voltage equations.

The formulation is not complete without a second mesh equation to provide for the decay of current in the phase which has just switched off. In this case it is transistor 5 which has just switched off (Fig. 4.6(a)) and transistor 1 which has just switched on.

Transistor 6 has been conducting continuously for over 60°. The current in phase C cannot decay to zero instantaneously at the commutation instant, no matter how fast the switching of the semiconductors, because of its inductance $L$. Current continues to flow in phase C and can be considered as mesh current that flows through diode 2, phase winding C, phase winding B, transistor 6, and back through diode 2. This mesh bypasses the supply and is in fact a freewheeling loop for which the voltage equation is

$$\frac{di_2}{dt} = \frac{1}{2L}[-V_Q - V_D - 2Ri_2 - (e_c - e_b)]$$

where $i_2 = i_c = -i_b$. In the absence of the driving voltage $V$, the value of this derivative is normally large and negative, and the current $i_2$ quickly decays to zero. If this happens before the next commutation, then it ceases to be necessary to integrate this equation. At the next commutation instant the circuit topology changes and the final values of $i_1$ and $i_2$ become the initial values of the next connection. The program must keep track of the commutation of current from one branch to another as the transistors switch on and off.

The effect of chopping can be approximately represented by considering that if the chopping frequency is high enough, then voltage p.w.m. with a duty-cycle of $d$ has the effect of reducing the supply voltage from $V$ to $Vd$. If the actual chopping is to be modelled, then the precise current-regulator or voltage-regulator algorithm must be built into the simulation and a smaller timestep must be used.

This technique can be incorporated with finite-element analysis by using the numerical field solutions to generate waveforms of flux-linkage vs. rotor position, from which the e.m.f.s can be derived after multiplying by $\omega$; and to calculate the inductance $L$. If saturation is not too great a problem, it is possible to perform the simulations fast enough so that in a single CAD program the motor parameters $e(\theta)$, $L$, and $R$ can be computed for given geometry and winding design, and the simulation run interactively and iteratively to optimize the design of the whole drive in one operation, i.e., with regard to the motor and controller design together.

## Problems for Chapter 4

1. An electric motor contains coils and magnets and the flux is fixed in magnitude. Can the flux-linkage of any coil vary?

2. A brushless d.c. motor has the cross-section shown in Fig. 4.15 with $r_1 = 30$ mm, $l_m = 8$ mm, $g = 0.8$ mm, and axial length $l = 35$ mm. Slotting is neglected and a single full-pitch stator coil is shown with 30 turns. The ceramic magnet has $B_r = 0.35$ T, $\mu_{rec} = 1$, and $H_c = 278$ kA/m. The current in the stator coil is zero.
   (a) Sketch the magnetic field set up by the magnet, by drawing 10 flux lines.
   (b) Sketch the variation of the radial component of airgap flux-density ($B_r$) around the inside of the stator, from 0 to 360°.
   (c) Draw an equivalent magnetic circuit and use it to calculate the flux crossing the

86  SQUAREWAVE PM BRUSHLESS MOTOR DRIVES

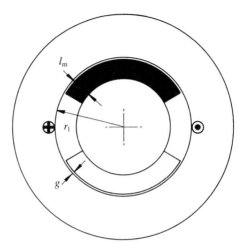

FIG. 4.15. Brushless d.c. motor cross-section.

airgap under each magnet pole. Neglect fringing and leakage flux, and assume that the permeability of the steel in the rotor and stator is infinite.
(d) Calculate $B_g$ in the airgap at the centre of the magnet arc, i.e. on the direct axis.
(e) Determine the m.m.f. across the magnet and the internal magnetizing force $H_m$.
(f) Calculate the flux-linkage of the stator coil in the position shown.
(g) Sketch the waveform of the coil flux-linkage as the rotor rotates through 360°.
(h) Determine the waveform and the peak value of the e.m.f. induced in the stator coil if the rotor rotates at 4000 r.p.m.
(i) Estimate the inductance of the stator coil.
(j) If the magnets require a magnetizing force of 1600 kA/m to magnetize them, estimate the coil current required to magnetize them in place. Comment on the feasibility of in-place magnetization.
(k) If the current in the stator coil is maintained constant at 5 A, determine the mechanical work that is done by the rotor in rotating 180° from the position shown.

3. (a) A PM brushless d.c. motor has a torque constant of 0.12 N m/A referred to the d.c. supply. Estimate its no-load speed in r.p.m. when connected to a 48 V d.c supply.
   (b) If the armature resistance is 0.15 Ω/phase and the total voltage drop in the controller transistors is 2 V, determine the stall current and the stall torque.
   (c) The d.c. current is 8.2 A when the motor is delivering 330 W of mechanical power to a load at 3400 r.p.m. The motor is star-connected and has two phases on at any instant, with a total of 2 V dropped across the two conducting transistors in series; this voltage drop can be assumed constant. The friction torque has been separately measured as 0.046 N m at this speed. If the supply voltage is 48 V d.c., calculate the efficiency of the complete drive and the separate power loss components due to (a) voltage drop in the transistors; (b) winding resistance; (c) friction; and (d) iron loss. If the iron loss is modelled by

means of a resistor connected in parallel with each phase of the motor, determine the value of this resistance.

4. A permanent-magnet d.c. commutator motor has a no-load speed of 6000 r.p.m. when connected to a 120 V supply. The armature resistance is 2.5 Ω and rotational and iron losses may be neglected. Determine the speed when the supply voltage is 60 V and the torque is 0.5 N m.

5. (a) Figure 4.16 shows a 'voice coil' actuator used in a Winchester disk drive for personal computers. The magnet length is 6 mm and the airgap is 2 mm. If the magnet is ferrite with a remanent flux-density of 0.3 T, calculate the flux-density in the airgap, assuming infinitely permeable iron components. Neglect fringing and leakage.

(b) The magnetic flux path is completed by a soft iron bridge. This is encircled by a coil of 200 turns mounted on the rotating 'armature'. If the current in the coil is 10 mA, calculate the torque. The radial distance from the centreline to the airgap is 25 mm and the height of the active magnetic parts is 50 mm top-to-bottom.

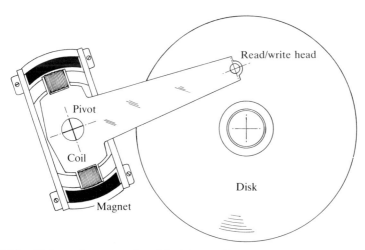

FIG. 4.16. Voice-coil actuator for disk drive read/write head.

6. (a) A permanent-magnet d.c. commutator motor has a stall torque of 1 N m with a stall current of 5 A. Estimate its no-load speed in r.p.m. when fed from a 28 V d.c. voltage supply.

(b) During an overload the motor temperature becomes excessive and the magnets lose 12 per cent of their remanent flux-density. If the armature resistance is 0.8 Ω, determine the speed at which the motor will run when the load torque is 0.3 N m. Assume a total brush voltage drop of 2 V and a supply voltage of 28 V d.c. Ignore friction and other losses.

# 5 Sinewave permanent-magnet brushless motor drives

The characteristic features of the brushless d.c. motors considered in Chapter 4 were:

(1) rectangular distribution of magnet flux in the airgap;
(2) rectangular current waveforms;
(3) concentrated stator windings.

The sinewave motor differs in all three respects. It has:

(1) sinusoidal or quasi-sinusoidal distribution of magnet flux in the airgap;
(2) sinusoidal or quasi-sinusoidal current waveforms;
(3) quasi-sinusoidal distribution of stator conductors; i.e. short-pitched and distributed or concentric stator windings.

The quasi-sinusoidal distribution of magnet flux around the airgap is achieved by tapering the magnet thickness at the pole edges and by using a shorter magnet pole arc, typically 120°. The quasi-sinusoidal current waveforms are achieved through the use of p.w.m. inverters which may be 'current regulated' to produce the best possible approximation to a pure sinewave. The use of short-pitched, distributed or concentric windings is exactly the same as in a.c. motors.

Indeed the sinewave motor is a simple synchronous motor. It has a rotating stator m.m.f. wave and can therefore be analysed with a phasor diagram; this is especially useful in designing the control system and calculating the performance.

Section 5.1 describes the ideal sinewave motor, that is, one with flux and winding distributions that are perfectly sinusoidal. We determine expressions for the torque, the open-circuit phase e.m.f. due to the magnet, the actual winding inductance, and the synchronous reactance. These expressions are derived in terms of the number of sine-distributed turns per pole.

In Section 5.2 these results are modified for practical windings by means of the standard winding factors of a.c. machines. These results provide the basis for the phasor diagram, which is developed in Section 5.3. In Section 5.4 the phasor diagram is used to develop the circle diagram and study its variation with speed; from this the torque/speed characteristic is derived. It is shown that the surface-magnet sinewave motor has limited capability to operate along a constant-power locus at high speed.

Section 5.5 presents a theoretical comparison of the torque per ampere and kVA requirements of squarewave and sinewave motors. Section 5.6 provides a comparison of wound-field and PM motors in such a way as to justify the

# IDEAL SINEWAVE MOTOR

preference for PM motors in smaller sizes. In Section 5.7 there is a treatment of the slotless motor, in which interest has revived as a result of the development of extremely high-energy magnet materials. Finally in Section 5.8 the subject of torque ripple is reviewed.

## 5.1 Ideal sinewave motor: torque, e.m.f., and reactance

### 5.1.1 Torque

The torque production can be analysed in terms of the interaction of the magnet flux and the stator ampere-conductor distribution, as in the squarewave motor. Figure 5.1 shows the basic concept for a machine with two poles. The stator conductors are distributed as a sine-distributed conductor-density around the stator bore, such that in any angle $d\theta$ the number of conductors is

$$\frac{N_s}{2} \sin p\theta \, d\theta.$$

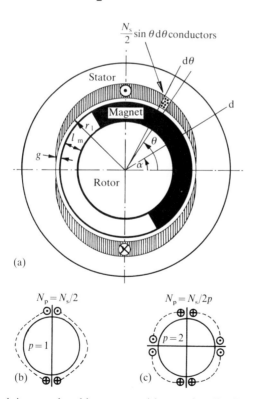

FIG. 5.1. (a) Ideal sinewave brushless motor with pure sine-distributed phase winding and permanent-magnet rotor with sine-distributed flux. (b) Turns/pole for two-pole winding. (c) Turns/pole for four-pole winding.

By integrating this expression over one-half of an electrical pole-pitch, i.e., from 0 to $\pi/2p$, and noting that one turn comprises two conductors in series, the number of turns per pole, $N_p$, is determined to be

$$N_p = \frac{N_s}{2p}.$$

This means that if all the poles are connected in series, $N_s$ is automatically equal to the number of turns in series in the whole winding, i.e. the number of turns in series per phase. The poles will be assumed to be so connected unless it is stated otherwise. The notional distribution of the 'turns per pole' is shown in Fig. 5.1(b) and (c) for two-pole and four-pole motors. In Section 5.2 we shall determine the effective value of $N_s$ for practical windings in slots.

In the following analysis the angle $\theta$ is in mechanical degrees or radians.

The stator ampere-conductor distribution is a sine-distributed current sheet of the same form as the conductor distribution, such that in angle $d\theta$ the ampere-conductors flowing in the positive direction (out of the paper) are

$$i \frac{N_s}{2} \sin p\theta \, d\theta.$$

The rotor magnet flux distribution is centred on its north d-axis, which is displaced by a positive angle $\alpha$ from the axis of the stator winding:

$$B(\theta) = \hat{B} \cos(p\theta - \alpha).$$

Here $\alpha$ is expressed in electrical radians, which will prove convenient later. The force on the elementary group of ampere-conductors is in the circumferential direction and is

$$F = \hat{B}li \frac{N_s}{2} \sin p\theta \cos(p\theta - \alpha) \, d\theta.$$

Together with the corresponding force on the opposite element, this force produces a couple $2Fr_1$ on the stator. An equal and opposite couple acts on the rotor, and the total electromagnetic torque on the rotor is the integral of the elementary contributions over the whole airgap periphery: over $p$ pole-pairs

$$T = -p \int_0^{\pi/p} 2Fr_1 \, d\theta$$

$$= -2r_1 \hat{B}li \frac{N_s}{2} \int_0^{\pi} \sin \theta \cos(\theta - \alpha) \, d\theta$$

$$= -\frac{\pi r_1 \hat{B}li N_s}{2} \sin \alpha \ \text{Nm}.$$

Maximum positive torque is obtained with $\alpha = -\pi/2$; that is, with the rotor north d-axis 'lagging' 90 electrical degrees behind the axis of the stator ampere-conductor distribution.

This analysis has been carried out for a stationary stator winding. To produce a constant torque with the rotor rotating at a steady speed, the stator ampere-conductor distribution must be made to rotate in synchronism with the rotor. This is done exactly as in induction motors, by means of a polyphase winding supplied with polyphase balanced currents. The most common number of phases is three, but two-phase motors are sometimes built, and occasionally four, six, and nine-phase motors. With $p$ pole-pairs, if $N_p = N_s/2p$ is the number of sine-distributed turns per pole, the total number of turns per phase is $N_s$. If the phase current flows through all these in series, then with balanced sinewave currents and three phase windings whose axes are 120 electrical degrees apart, the rotating ampere-conductor distribution can be derived as

$$\hat{i} \cos \omega t \frac{N_s}{2} \sin p\theta + \hat{i} \cos\left(\omega t - \frac{2\pi}{3}\right) \frac{N_s}{2} \sin\left(p\theta - \frac{2\pi}{3}\right)$$
$$+ \hat{i} \cos\left(\omega t + \frac{2\pi}{3}\right) \frac{N_s}{2} \sin\left(p\theta + \frac{2\pi}{3}\right) = \frac{3}{2} I\sqrt{2} \frac{N_s}{2} \sin(p\theta - \omega t),$$

and the rotating magnet flux distribution is

$$B(\theta) = \hat{B} \cos(p\theta - \omega t - \alpha).$$

The torque is obtained using the same expression as for the stationary winding, with $p\theta - \omega t$ substituted for $p\theta$, thus;

$$T = \frac{3}{2} I\sqrt{2} \frac{\pi r_1 l \hat{B} N_s}{2} \sin \beta$$

where $\beta = -\alpha$. The angle $\beta$ is called the torque angle, and is positive for motoring; it is measured in electrical radians or degrees. If the phase winding is divided into '$a$' parallel paths, then the equation remains valid provided that $I$ is the total phase current and there is no change in the total number of turns per phase, i.e. $N_s$. (The number of turns in series per phase is then $N_s/a$.)

It is worth emphasizing that the flux-density $\hat{B}$ in the torque equation is the peak airgap flux-density produced by the magnet acting alone; in other words, it is the open-circuit value and does not include any contribution due to the m.m.f. of the stator currents. Although the armature-reaction m.m.f. modifies the airgap flux-density, it does not figure in the torque expression unless it significantly affects the saturation level of the magnetic circuit. Physically the stator may be regarded as being incapable of producing torque on itself. The armature reaction flux is aligned with the stator ampere-conductor distribution and therefore has an effective 'torque angle' of zero. The armature reaction flux does, however, rotate and induce a voltage drop in the phase windings, which must be overcome by the supply voltage. (See Section 5.2).

In normal operation the stator frequency (in rad/sec) is made equal to the rotation frequency (in electrical rad/sec), i.e.,

$$\omega = 2\pi f \text{ electrical rad/sec}$$

and the mechanical angular velocity is

$$\omega_m = \frac{\omega}{p}.$$

The stator ampere-conductor distribution rotates in synchronism with the rotor and the torque angle is kept constant, usually by means of a simple form of 'vector' control or 'field-oriented control' which requires a shaft position sensor (i.e. encoder or resolver feedback). If the supply frequency and the rotation frequency were unequal, the motor would be running asynchronously. No average torque would be produced, but there would be a large alternating torque at the 'beat' frequency or pole-slipping frequency.

### 5.1.2 E.m.f.

The e.m.f. equation of the sinewave motor can be derived by considering the e.m.f. induced in the elementary group of conductors in Fig. 5.1. Noting that Fig. 5.1 is drawn for a two-pole machine ($p=1$), for a machine with $p$ pole-pairs in series this e.m.f. is

$$de = B(\theta) l \omega_m r_1 \frac{N_s}{2} \sin p\theta \, d\theta.$$

But

$$B(\theta) = \hat{B} \cos(p\theta - \omega t - \alpha)$$

and by integrating the contributions of all the elementary groups of conductors we get the instantaneous phase e.m.f.:

$$e = 2p \int_0^{\pi/p} de = \frac{\hat{B} l \omega r_1 N_s \pi}{2p} \sin(\omega t + \alpha).$$

The r.m.s. phase e.m.f. is therefore

$$E_{ph} = \frac{\pi}{2\sqrt{2}} \frac{\hat{B} l \omega r_1 N_s}{p} \text{ V}_{r.m.s.}$$

and the line-line e.m.f. is $E\sqrt{3}$. In Section 5.2 this equation is used to derive the fundamental e.m.f. for a practical winding in slots, and practical expressions for $N_s$ are derived.

It is worth emphasizing again that the flux-density $\hat{B}$ is the peak airgap flux-density produced by the magnet acting alone; in other words, it is the open-circuit value and does not include any contribution due to the m.m.f. of the

# IDEAL SINEWAVE MOTOR

stator currents. The voltage drop induced by flux attributable to armature reaction is dealt with below.

The e.m.f. equation can also be derived from Faraday's law. This alternative method is included here because it is the basis of the phasor diagram and provides the means for calculating the inductive volt drop due to armature reaction. Faraday's law is more rigorous than the BLV formulation, but it is useful to show that for $E$ both methods give the same result.

By Faraday's law, the instantaneous e.m.f. induced in the stationary phase winding of Fig. 5.1 is given by

$$e = -\frac{d\psi}{dt} \text{ V}$$

where $\psi$ is the instantaneous flux-linkage. To calculate the flux-linkage consider the coil formed by the elementary group of conductors within the angle $d\theta$ at angle $\theta$, and assume that the return conductors of this coil are located within the angle $d\theta$ at angle $-\theta$. Although Fig. 5.1 is drawn for a two-pole machine ($p=1$), the results are derived for $p$ pole-pairs.

On open-circuit there is no current in the coil, and all the flux is due to the magnet. The flux through the elementary coil is

$$\phi = \int_{-\theta}^{\theta} B(\theta) r_1 l \, d\theta \text{ Wb}.$$

But

$$B(\theta) = \hat{B} \cos(p\theta - \omega t - \alpha)$$

so the integral gives

$$\phi = \frac{\hat{B}Dl}{p} \sin p\theta \cos(\omega t + \alpha) \text{ Wb}$$

where $D = 2r_1$ is the stator bore. The flux per pole can be extracted from this expression by setting $\theta = \pi/p$ and $t = 0$. Thus

$$\Phi_M = \frac{\hat{B}Dl}{p} \text{ Wb}.$$

This is a fixed flux that rotates with the rotor. The flux-linkage of the elementary coil is

$$d\psi = \phi \left[ \frac{N_s}{2} \sin p\theta \, d\theta \right] \text{ V s}.$$

The total flux-linkage of the winding is obtained by integrating the contributions of all the elementary coils: with $p$ pole-pairs the result is

$$\psi_M = p \int_0^{\pi/p} d\psi = \hat{\psi}_M \cos(\omega t + \alpha) \text{ V s}$$

where

$$\hat{\psi}_M = \frac{\hat{B} l r_1 N_s \pi}{2p} = \frac{\pi}{4} N_s \Phi_M \text{ V s.}$$

The subscript 'M' has been added as a reminder that the flux is produced only by the magnet. By Faraday's law the instantaneous phase e.m.f. is

$$e = -\frac{d\psi_M}{dt} = \omega \hat{\psi}_M \sin(\omega t + \alpha) \text{ V.}$$

The r.m.s. phase e.mf. is

$$E_{ph} = \frac{\omega \hat{\psi}_M}{\sqrt{2}} = \frac{\pi}{2\sqrt{2}} \frac{\hat{B} l \omega r_1 N_s}{p} \text{ V}_{r.m.s.}$$

as before.

### 5.1.3 Inductance of phase winding

Figure 5.2 shows the same single-phase, two-pole sine-distributed winding as in Fig. 5.1. But the flux is now produced by the current in the stator winding, and we assume that the magnet is unmagnetized while we calculate the inductance by determining the flux-linkage of the winding due to its own current, $i$. If the steel in the rotor and stator is assumed to be infinitely permeable, then the m.m.f. is concentrated entirely across the two airgaps. Across each airgap the m.m.f. drop is equal to one-half the ampere-conductors enclosed within an 'Ampere's law contour' or flux-line:

$$F_g = H_g g'' = \frac{1}{2} \int_\theta^{\pi/p - \theta} i \frac{N_s}{2} \sin p\theta \, d\theta$$

$$= \frac{N_s i}{2p} \cos p\theta.$$

The flux-density across the gap and the magnet is assumed to be radial, and the magnet is assumed to be equivalent to an airgap of length $l_m/\mu_{rec}$, as before: this gives an 'effective airgap'

$$g'' = g' + \frac{l_m}{\mu_{rec}}.$$

Hence

$$B(\theta) = \mu_0 H_g = \frac{\mu_0 N_s i}{2pg''} \cos p\theta = \hat{B}_a \cos p\theta.$$

The subscript 'a' has been added to the peak airgap flux-density to denote that it is generated by armature current.

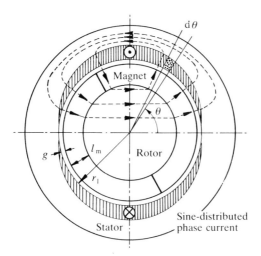

FIG. 5.2. Calculation of armature reaction m.m.f. of sinewave winding.

By integrating the flux-density around the periphery of the airgap, the fundamental armature-reaction flux per pole can be determined as

$$\Phi_a = \frac{\hat{B}_a D l}{p} \text{ Wb}$$

where, again, $D = 2r_1$ is the stator bore. This expression has exactly the same form as the flux per pole of the magnet and therefore it produces a flux-linkage

$$\psi_a = \frac{\pi}{4} N_s \Phi_a.$$

The self-inductance is obtained as the flux-linkage per ampere. With $N_s$ turns in series per phase

$$L_g = \frac{\pi \mu_0 N_s^2 l r_1}{4 p^2 g''} \text{ H}.$$

The inductance is only half the value which would be obtained with the same number of turns concentrated into one pair of slots spanning 180 electrical degrees (Section 4.7).

The inductance calculated above is the actual 'airgap' inductance, i.e. the value which would be measured with the rotor stationary and unmagnetized; with the other phases open-circuited; and with negligible leakage inductance from the slots or the end-turns. The actual inductances of three simple full-pitch windings with one, two, and three slots per pole per phase were calculated in Section 4.7 on the same basis. The sine-distributed winding is a hypothetical case with an infinite number of slots per pole per phase, the number of conductors per slot being modulated sinusoidally to produce a sine-distributed airgap m.m.f.

### 5.1.4 Synchronous reactance

Earlier we showed that three sine-distributed phase windings carrying balanced three-phase sinusoidal currents produce a sine-distributed ampere-conductor distribution represented by the expression

$$\frac{3}{2} I \sqrt{2} \frac{N_s}{2} \sin(p\theta - \omega t).$$

This sets up a rotating flux wave

$$\hat{B}_a \sin(p\theta - \omega t)$$

where

$$\hat{B}_a = \frac{\mu_0}{g''} \frac{3}{2} I \sqrt{2} \frac{N_s}{2p}.$$

This rotating flux wave, established by armature reaction, generates voltages in all three phases. In each phase the voltage is proportional to $I$ and is therefore regarded as the voltage drop $X_s I$ across a fictitious reactance, the 'synchronous reactance', $X_s$. By substituting the peak flux-density into the expression derived earlier for e.m.f., and dividing by $I$, we get

$$X_s = \frac{3\pi\mu_0 N_s^2 l r_1 \omega}{8p^2 g''} \text{ Ohms}.$$

This expression applies to an ideal two-pole sine-distributed three-phase winding with $N_s$ turns in series per phase, and it neglects the leakage inductance of the slots and end-turns. To obtain a practical formula for a real winding, we must first find an effective value for the sine-distributed turns. This is done by means of Fourier analysis and winding factors in the next section.

## 5.2 Sinewave motor with practical windings

The expressions developed for torque, e.m.f., inductance, and reactance in Section 5.1 can be modified for practical windings by means of Fourier analysis and harmonic winding factors. It is normally assumed for a.c. machines that the operation is dominated by the fundamental space-harmonic components of the conductor and flux distributions. Machines are designed to operate in this way to minimize the torque ripple and parasitic losses that would be caused by nonsynchronous rotating field components. When the

# SINEWAVE MOTOR WITH PRACTICAL WINDINGS

harmonic effects are neglected in favour of the fundamentals, the effective number of sine-distributed turns is easy to derive, and can be substituted in the equations of Section 5.1.

Consider first the concentrated two-pole full-pitched coil in Fig. 5.3, with one slot per pole per phase and $N$ conductors in each slot. This winding is regarded as having $N/2$ turns per pole, which may seem a little artificial; however, this makes the treatment of the two-pole machine consistent with that of windings with more than two poles, where the coils can usually be identified clearly with the poles (see Fig. 5.1(b)). The rectangular m.m.f. wave has a fundamental harmonic component equal to $4/\pi$ times the peak flat-topped value. The effective number of sine-distributed turns per pole is therefore

$$N_s = \frac{4}{\pi} N.$$

If there are $p$ pole-pairs in series, the actual number of series turns per phase is

$$N_{ph} = 2pN.$$

The effective number of sine-distributed turns in series per phase is $2p$ times the number of sine-distributed turns per pole, i.e. $N_s$. Therefore

$$N_s = \frac{4}{\pi} N_{ph}.$$

The corresponding result for any winding distributed in slots can be built up by considering the winding as an aggregate of coils, each of which occupies one

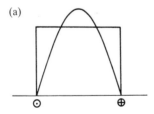

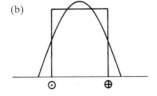

FIG. 5.3. Distribution and fundamental harmonic component of single full-pitched coil.

slot pair. For a distributed full-pitch winding with $q$ slots per pole per phase, such as those in Section 4.7, the effective sine-distributed turns are given by

$$N_s = \frac{4}{\pi} k_{d1} N_{ph}$$

where $k_{d1}$ is the 'distribution' or 'spread' factor for the fundamental, and is given by

$$k_{d1} = \frac{\sin\frac{q\gamma}{2}}{q \sin\frac{\gamma}{2}}.$$

Here $\gamma$ is the slot-pitch in electrical degrees and $q$ is the number of slots per pole per phase. With three phases,

$$\gamma = \frac{\pi}{3q}.$$

For example, in a four-pole three-phase machine with 2 slots per pole per phase there are 24 slots and the fundamental distribution factor is 0.966. With 3 slots per pole per phase, there are 36 slots and the fundamental distribution factor is 0.960. The distribution factor reduces the actual number of turns per phase to the number of turns per phase required in a concentrated coil that will produce the same fundamental component of airgap m.m.f. with the same current flowing.

Many windings are not only distributed into more than one slot per pole per phase, but are also 'chorded' or short-pitched. For a single chorded coil the fundamental component of airgap m.m.f. is equal to that produced by a full-pitched coil with $k_{p1}$ times the number of turns, where $k_{p1}$ is the fundamental 'pitch factor' or 'chording factor':

$$k_{p1} = \cos\frac{\varepsilon}{2}.$$

For example, in a winding with six slots per pole, a winding with coils spanning five slots would have 5/6 pitch and

$$\varepsilon = \frac{\pi}{6}$$

giving

$$k_{p1} = 0.966.$$

In the case of skewed windings there is a skew factor that is treated in the same way as the distribution and pitch factors; this is given by

# SINEWAVE MOTOR WITH PRACTICAL WINDINGS

$$k_{s1} = \frac{\sin \sigma}{\sigma}$$

where $\sigma$ is the half the skew angle in mechanical radians or degrees, (Levi 1984).

When a winding has coils that are distributed, chorded, and skewed, the overall result is a 'fundamental winding factor'

$$k_{w1} = k_{d1} k_{p1} k_{s1}$$

and the number of effective sine-distributed turns per phase is given by

$$N_s = \frac{4}{\pi} k_{w1} N_{ph}$$

where $N_{ph}$ is the actual number of turns in series per phase.

Substitution of this result in the equations of Section 5.1 produces the following formulas for the torque, e.m.f., and synchronous reactance.

The open-circuit phase e.m.f. is

$$E_{ph} = \frac{2\pi}{\sqrt{2}} (k_{w1} N_{ph}) \Phi_{M1} f \; V_{r.m.s.}$$

The subscript '1' has been added to the magnet flux per pole to denote its fundamental space-harmonic component. If, as is usually the case, the magnet flux distribution is not perfectly sinusoidal, the fundamental component should be used. The winding has the effect of filtering out the harmonics in the flux distribution. The degree of suppression of individual harmonics is determined by the harmonic winding factors.

The e.m.f. can be written in an alternative useful form by recognizing that the peak phase flux-linkage due to the magnet is

$$\hat{\psi}_{M1} = k_{w1} N_{ph} \Phi_{M1}.$$

With this, the e.m.f. can be written

$$E_{ph} = \frac{1}{\sqrt{2}} \omega \hat{\psi}_{M1} = E_q.$$

The subscript 'q' means that the e.m.f. phasor is aligned with the q-axis in the phasor diagram (see Section 5.3), since it leads the flux-linkage by 90 electrical degrees.

From Section 5.1 the torque is given by

$$T = \frac{3}{2} I \sqrt{2} \, \frac{\pi r_1 \hat{B}_{M1} l N_s}{2} \sin \beta$$

$$= \frac{3\pi p}{8} I \sqrt{2} \, \Phi_{M1} N_s \sin \beta.$$

Replacing $N_s$ by $(4/\pi)k_{w1}N_{ph}$ and using the above expressions for flux-linkage and $E_{ph}$, the result is

$$T = \frac{3p}{\omega} E_q I \sin \beta \text{ N m}.$$

Turning now to the actual phase inductance, it is meaningless to substitute the effective sine-distributed turns. The actual inductance must include all the self flux-linkage, not just the fundamental component. The inductance formula for a practical winding therefore remains the same as in Section 4.7, including the leakage inductance.

The synchronous inductance or reactance is quite different from the actual inductance, being associated with the voltage drop in one phase caused by the fundamental component of rotating armature-reaction flux, under balanced conditions with all three phases in operation. Substituting for the effective sine-distributed turns per phase, we get

$$X_{sg} = \frac{6\mu_0 Dlf}{p^2 g''} (k_{w1} N_{ph})^2 \text{ Ohms}.$$

To this value must be added the per-phase leakage reactance, giving the total synchronous reactance

$$X_s = X_{sg} + X_\sigma.$$

## 5.3 Phasor diagram

If the magnets are on the rotor surface, and if the shaft cross-section is circular, the sinewave motor is a 'non-salient pole' synchronous machine: that is, its d-axis and q-axis synchronous reactances are equal. The calculation of the synchronous reactance was discussed in Section 5.2. In the steady state with balanced, sinusoidal phase currents the operation can be represented by the phasor diagram shown in Fig. 5.4.

The construction of the phasor diagram can be understood by considering in turn how each phasor in it is derived. This analysis provides the basis for understanding the performance characteristics, and what happens at different speeds and torques. It is also the basis for the design of effective controls.

The squarewave motor (Chapter 4) is not amenable to phasor analysis because the stator ampere-conductor distribution does not rotate: it is not a rotating-field machine in the a.c. sense.

In Section 5.2 it is shown that the open-circuit phase e.m.f. $E$, is a sinewave which lags behind the magnet flux by 90 electrical degrees. Its r.m.s. value is

$$E_{ph} = \frac{2\pi}{\sqrt{2}} k_{w1} N_{ph} \Phi_{M1} f \text{ V}_{r.m.s.}.$$

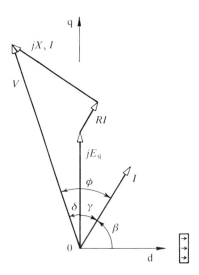

FIG. 5.4. Phasor diagram of surface-magnet sinewave motor.

The phasor relationship between $E_{ph}$ and the fundamental flux-linkage due to the magnet can be written as follows:

$$\tilde{E}_{ph} = jE_q = j\omega\tilde{\psi}_{Md1} \text{ V}.$$

The subscript 'd' in the magnet flux-linkage denotes that this phasor is synchronized with the direct axis of the rotor. The subscript '1' indicates the fundamental space-harmonic component, whose phasor rotates on paper in synchronism with the rotor in physical space. The subscript 'q' in $E_q$ denotes that it is synchronized with the q-axis. In all phasor diagrams the d-axis is taken to define the real axis of the complex plane of the phasor diagram, as in Fig 5.4. In physical space the rotor flux axis is the direct axis, i.e., the centre of the magnet pole arc. The rotation is 'cancelled out' by choosing this as a reference axis, so that an 'observer' rotates with this frame of reference at synchronous speed, and perceives a magnetic field that is stationary with respect to the rotor. It can be shown by mathematical transformation (Park's transformation) that this choice of reference frame is equivalent to cancelling out the $e^{j\omega t}$ from the voltages and currents, leaving a phasor diagram that is stationary on paper. Note that as the rotor *flux* rotates in space, the flux-*linkage* of a stationary phase winding pulsates in time.

The angle $\alpha$ defines the position of the physical rotor d-axis at $t=0$, measured from the positive axis of the phase 'a' winding.

Next we determine the current phasor. To preserve the relationship between phasors and rotating fields, we need the rotating ampere-conductor distribution of all three phases taken together. Assuming balanced currents,

$$i_a = \hat{i} \cos \omega t$$
$$i_b = \hat{i} \cos(\omega t - 2\pi/3)$$
$$i_c = \hat{i} \cos(\omega t + 2\pi/3).$$

The axes of the three windings are displaced by 120 electrical degrees in such a way that the rotor d-axis rotates past them in the sequence a, b, c. This gives rise to an ampere-conductor distribution

$$\hat{i} \cos \omega t \, \frac{N_s}{2} \sin \theta + \hat{i} \cos\left(\omega t - \frac{2\pi}{3}\right) \frac{N_s}{2} \sin\left(\omega t - \frac{2\pi}{3}\right)$$
$$+ \hat{i} \cos\left(\omega t + \frac{2\pi}{3}\right) \frac{N_s}{2} \sin\left(\theta + \frac{2\pi}{3}\right)$$
$$= \frac{3}{2} I \sqrt{2} \, \frac{N_s}{2} \sin(\theta - \omega t).$$

The value of $N_s$ for practical windings is discussed in Section 5.2. The r.m.s. current in phase 'a' is $I$ amperes, and its phase relative to that of the open-circuit flux-linkage of phase 'a' is evidently lagging by the angle $\alpha$. Correspondingly, the rotating ampere-conductor distribution lags the rotor d-axis and the flux by the same angle (measured in electrical degrees). We have noted in Section 5.1 that for positive motoring torque the angle $\alpha$ must be negative, and we substituted the angle $\beta = -\alpha$ which for motoring is positive. This is shown in Fig. 5.4. In motoring the axis of the ampere-conductor distribution leads the rotor d-axis by the angle $\beta$, as though the ampere-conductors were dragging the rotor round behind them.

The phasor $R\tilde{I}$ represents the voltage drop across the phase resistance, and is clearly parallel to $\tilde{I}$. Similarly, the voltage drop across the synchronous reactance is represented by $jX_s\tilde{I}$, and leads the current phasor by 90°. The sum of the back-e.m.f. and voltage drop phasors must be equal to the applied voltage at the terminals. Thus

$$\tilde{V}_{ph} = \tilde{E} + R\tilde{I} + jX_s\tilde{I}.$$

Figure 5.5 shows the phasor diagram with the current and voltages resolved into d- and q-axis components. In both cases the resistance is neglected in order to bring out the essential 'mechanism' of operation: obviously the resistance must be considered in practice, but it does not exert the controlling influence. In Fig. 5.5(a) the current leads the d-axis by an angle less than 90°, and lags the q-axis by the angle $\gamma$, where

$$-\gamma = \frac{\pi}{2} - \beta.$$

The minus sign simply means that $\gamma$ is measured from the q-axis in the normal positive (anticlockwise) direction. The convenience of this will appear later.

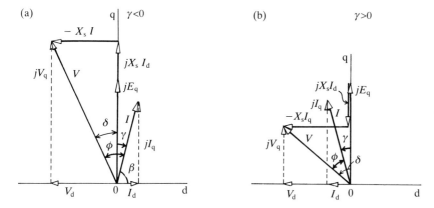

FIG. 5.5. Phasor diagram with resistance neglected, and with current and voltages resolved into d- and q-axis components: (a) 'magnetizing' d-axis current; (b) 'demagnetizing' d-axis current.

The d-axis and q-axis components of current are given by

$$I_d = -I \sin \gamma$$
$$I_q = I \cos \gamma.$$

If $I_d$ is positive, as in Fig. 5.5(a), the armature or stator current produces an m.m.f. distribution around the airgap that tends to increase the d-axis flux produced by the magnet. In this condition the armature reaction is said to be 'magnetizing'. The flux produced by the stator m.m.f. induces the voltages $jX_s I_d$ in the q-axis, and $-X_s I_q$ in the d-axis of the phasor diagram. The q-axis component adds to $jE_q$. The magnet flux-density is increased, so that the operating point moves up the demagnetization characteristic and may end up in the first quadrant with $B_m > B_r$.

In Fig. 5.5(b), $I_d$ is negative and the stator m.m.f. is negative and 'demagnetizing'. The power factor angle is given by

$$\phi = \delta - \gamma$$

and this is clearly less than in Fig. 5.5(a). In other words, a high power factor is associated with operation in the 'demagnetizing' mode. This also means that a highly coercive magnet permits high power factor and reduces the kVA requirement in the converter.

## 5.4 Sinewave motor: circle diagram and torque/speed characteristic

The phasor diagram and associated voltage equations can be used to derive control laws and predict the performance of the sinewave motor in closed analytical form. Apart from the simple relationships derived in Section 4.4, these results are not available for the squarewave motor other than by computer simulation.

If resistance is neglected, then from Fig. 5.5

$$V_d = -V \sin \delta = -X_s I_q$$
$$V_q = V \cos \delta = E_q + X_s I_d.$$

The electromagnetic torque is given by the following equation for synchronous machines which is derived in all the classical texts on d, q-axis theory:

$$T = 3p[\Psi_d I_q - \Psi_q I_d]$$

where $p$ is the number of pole-pairs, and the phase number is three. The flux-linkages, $\Psi_d$ and $\Psi_q$, are r.m.s. per-phase values given by

$$\omega \Psi_d = E_q + X_d I_d$$
$$\omega \Psi_q = X_q I_q.$$

Therefore the torque is given by

$$T = \frac{3p}{\omega}[E_q I_q + (X_d - X_q)I_d I_q] \text{ N m}.$$

The first term is called 'magnet alignment torque' and the second term is called 'reluctance torque'. In surface-magnet motors,

$$X_d = X_q = X_s.$$

There is no reluctance torque and therefore

$$T = \frac{3p}{\omega} E_q I_q \text{ N m}.$$

This equation is valid even when the resistance is not zero. A contour of constant torque is a horizontal straight line in the phasor diagram of the current $\hat{I} = I_d + jI_q$.

At a given speed, $E_q$ is fixed by the magnet flux and the torque is proportional to the q-axis current $I_q$. Since $E_q$ is itself proportional to speed, this relationship is valid also at zero speed. The linear relationship between torque and current is an important feature: it simplifies the controller design and makes the dynamic performance more regular and predictable. The same property is shared by the squarewave motor and the PM d.c. commutator motor.

The amount of current that can be supplied is limited by the heat dissipation capability of the motor, and by the current rating of the converter. It is also limited by the converter voltage, which must overcome both the back-e.m.f. and the voltage drops across the synchronous reactance and the resistance. In the complex plane of the phasor diagram the maximum continuous converter current $I_c$ limits the current phasor to a circular locus described by the equation

$$I_d^2 + I_q^2 = I_c^2.$$

This is shown in Fig. 5.6. The maximum converter voltage $V_c$ limits the current phasor to another circular locus with a different radius and a different centre; this is derived as follows: if

$$V_d^2 + V_q^2 = V_c^2$$

then

$$X_s^2 I_q^2 + (E_q + X_s I_d)^2 = V_c^2$$

and

$$I_q^2 + \left[ I_d + \frac{E_q}{X_s} \right]^2 = \frac{V_c^2}{X_s^2}$$

which represents the circle in Fig. 5.6 with centre at the point $(-E_q/X_s, 0)$. The offset is independent of frequency and speed since both $E_q$ and $X_s$ are proportional to frequency. If the controller is a p.w.m. sinewave supply, then at low speeds there is usually sufficient voltage available to cause the voltage-limited locus to enclose the current-limited locus, so that maximum current can be obtained at any angle. The most advantageous angle for the current is obviously along the q-axis, since this maximizes the torque per ampere. In

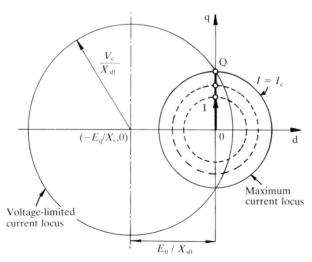

FIG. 5.6. Circle diagram of sinewave PM motor showing current-limit and voltage-limit loci.

general at low speeds the p.w.m. duty cycle is low and the phase voltage is 'chopped down' to a value much less than $V_c$. Operation is along 0Q, with torque proportional to current.

As the speed and frequency increase, the current-limit locus remains fixed, but there comes a speed at which the radius of the voltage-limit locus begins to decrease. This happens when the p.w.m. duty cycle reaches its maximum, and the phase voltage equals the maximum available sinewave voltage $V_c$ from the converter. The p.w.m. control is sometimes said to have 'saturated' at this point, which is shown in heavy line in Fig. 5.7. Operation along 0Q is still possible, but it is only just possible for the current to reach its rated value $I_c$ at Q. The speed at which this happens is called the 'corner-point' speed. It is the maximum speed at which full torque can be developed.

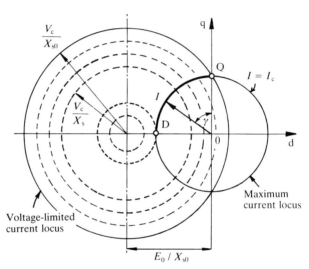

FIG. 5.7. Circle diagram of sinewave PM motor showing effect of speed on the voltage-limited current locus. The heavy line corresponds to the 'corner point' at which the p.w.m. control has just saturated.

If the speed increases further, the radius of the voltage-limit locus decreases. The maximum current is now at the intersection of the two circles. Although it is still possible to get the magnitude of the current up to $I_c$, it is not possible to orient a current of this magnitude along the q-axis and therefore the torque decreases. The decreasing radius of the voltage-limit circle 'drags' the maximum current phasor further and further ahead of the q-axis, and the q-axis current decreases while the d-axis current increases in the negative (demagnetizing) direction. This continues until point D, at which speed the maximum current $I_c$ can still be forced into the motor, but only just, and it is entirely in the d-axis so that no torque is developed. The power factor at this

# CIRCLE DIAGRAM AND TORQUE/SPEED CHARACTERISTIC 107

point is zero and the current is wholly demagnetizing—an onerous operating point for the magnet, especially if the temperature is high.

The loci 0Q and 0D together form the limiting locus for the current phasor throughout the whole speed range, and give rise to the torque/speed characteristic of Fig. 5.8. Along 0Q maximum torque can be developed with maximum current $I_c$ oriented along the q-axis in the phasor diagram. In physical space in the motor, the axis of the rotating ampere-conductor distribution is 90° ahead of the rotor d-axis. Q is the corner-point, the maximum speed at which full torque can be developed. Along QD the torque decreases until at point D it is zero, with maximum current $I_c$ still flowing but oriented in the d-axis in the negative (demagnetizing) direction.

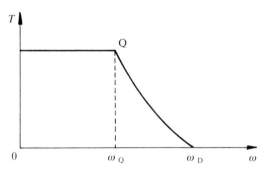

FIG. 5.8. Torque/speed characteristic of sinewave motor.

The ratio between the speeds at points D and Q is

$$k = \frac{\omega_D}{\omega_Q} = \frac{f_D}{f_Q}.$$

If we continue to neglect resistance, at Q we have

$$I_d = 0; \; I_q = I_c; \; V_q = E_0; \text{ and } V_d = -X_{s0}I_c.$$

The subscript 0 in the synchronous reactance denotes the value at the corner-point frequency. From the phasor diagram

$$\tilde{I} = jI_q = jI_c \text{ and } V_c^2 = E_0^2 + X_{s0}^2 I_c^2$$

so that

$$I_c = \sqrt{\frac{V_c^2 - E_0^2}{X_{s0}^2}}.$$

At point D,

$$\tilde{I} = I_d = -I_c = \frac{V_c - kE_0}{kX_{s0}}.$$

Equating the expressions for $I_c$ at the two speeds, we get the following expression for $k$:

$$k = \frac{V_c}{E_0 - \sqrt{V_c^2 - E_0^2}}.$$

Now define $e_0$ as the per-unit open-circuit voltage at the corner-point, with the maximum r.m.s. a.c. voltage of the converter as the base voltage:

$$e_0 = \frac{E_0}{V_c}.$$

Then

$$k = \frac{1}{e_0 - \sqrt{1 - e_0^2}}.$$

For example, suppose the motor is required to operate at maximum torque up to 3000 r.p.m., and be capable of just reaching 6000 r.p.m. with no load. Then $k = 2$, and to achieve this $e_0$ must be 0.911.

At the corner point the d-axis current is zero and, from the phasor diagram at this point, the power factor is given by

$$\cos\phi_0 = \frac{E_0}{V_c} = e_0.$$

Also,

$$\sin\phi_0 = \frac{X_q I_q}{V} = \frac{X_s I_c}{V_c} = x_{s0}.$$

Here $x_{s0}$ is the per-unit synchronous reactance at the corner-point frequency, that is, the synchronous reactance normalized to a base of $V_c/I_c$. In the example quoted above with $k = 2$, the corner-point power factor is 0.911 and the per-unit synchronous reactance is $x_{s0} = 0.411$. This is a large value for a surface-magnet motor: typically $x_{s0}$ is in the range 0.1–0.2, giving a higher corner-point power factor but a value of $k$ closer to 1.0. (Methods for calculating the reactance are discussed in section 5.1.4) The surface-magnet motor has very limited capability to operate above its corner-point speed, and certainly cannot maintain a constant-power characteristic as the speed is increased. The fundamental physical reason for this is that the airgap flux is fixed predominantly by the magnet and direct field-weakening is not possible to any useful degree. In this discussion the resistance has been neglected, but in practice it introduces further phase shifts which make the speed-torque envelope even more restricted than that presented. An analysis that includes resistance is given by Leonhard (1985).

If the speed is increased beyond point D, there is a risk of overcurrent because the back-e.m.f. $E_q$ continues to increase while the terminal voltage remains constant. The current is then almost a pure reactive current (with nearly zero power factor) flowing from the motor back to the supply. There is a small q-axis current and a small torque because of losses in the motor and the converter. The power flow is reversed and this mode of operation is possible only if the motor 'over-runs' the converter or is driven by an external load or prime-mover. The reactive current is limited only by the synchronous reactance, and as the speed increases it approaches the short-circuit current $E_q/X_s$, which may be many times the normal continuous rating of the motor windings or the converter, and may be sufficient to partially demagnetize the magnets, particularly if their temperature is high. The current is rectified by the flyback diodes in the converter and there is a risk not only of overcurrent in the diodes but also of overvoltage on the d.c. side of the converter, especially if a filter capacitor and a.c. line rectifier are used to supply the d.c. Fortunately this operating condition is unusual, but in any system design the possibility should be assessed. An effective solution is to use an overspeed relay to short-circuit the phase windings into a three-phase resistor or a short-circuit, to produce a braking torque without stressing the converter.

## 5.5 Torque per ampere and kVA/kW of squarewave and sinewave motors

With a torque angle of 90° the torque per r.m.s. ampere of phase current in the three-phase sinewave motor is

$$T/I = \frac{3}{2}\sqrt{2}\,\frac{\pi r_1 l \hat{B} N_{ph}}{2} \text{ N m/A}$$

and the torque per peak ampere is $\sqrt{2}$ times smaller. In the squarewave motor (assuming 180° magnet arcs, star connection, and 120° squarewave phase currents), the torque per peak ampere of phase current is

$$T/i = 4 r_1 \hat{B} l N_{ph} \text{ N m/A}.$$

The r.m.s. phase current $I$ is derived from the 120° squarewave as:

$$I = i\sqrt{\frac{2}{3}}$$

where $i$ is the peak or flat-top value of the phase current. Therefore the torque per r.m.s. ampere of phase current is

$$T/I = 4\sqrt{\frac{3}{2}}\,r_1 \hat{B} l N_{ph} \text{ N m/A}.$$

With these expressions the ratio of torque per ampere in the two motors can be compared. With equal r.m.s. phase currents, the torque of the squarewave motor exceeds that of the sinewave motor by the factor

$$\frac{4\sqrt{\frac{3}{2}}}{\frac{3}{2} \times \sqrt{2} \times \frac{\pi}{2}} = 1.47.$$

With equal peak currents, the factor is 1.27. The squarewave motor has a significantly higher torque per ampere. However, the comparison assumes equal peak magnet flux-densities in the airgap, which is likely to require significantly more magnet and a thicker stator yoke in the squarewave motor. To rectify this imbalance it is perhaps better to compare the motors on the basis of the same flux per pole. For the same peak flux-density the flux per pole of the square-wave motor exceeds that of the sinewave motor by the factor $\pi/2$. Now with equal r.m.s. phase currents the torque ratio is only 0.94, and with equal peak currents it is 0.81. This analysis neglects many important effects, such as armature reaction and losses, but it indicates that with equal amounts of copper, iron, and magnet, the torque per ampere is not greatly different between the two machines.

The comparison is now carried to the volt-ampere requirements of the electronic converter. A simple estimate of the converter 'rating' can be made in terms of the total kVA rating of its main switches, per kW of power fed to the motor. The relevant parameters are defined as follows. With respect to the r.m.s. current in each switch, if $q$ is the phase number,

$$\text{r.m.s. kVA/kW} = 2q \times I_s \times V_s$$

where $I_s$ is the r.m.s. current in each switch and $V_s$ is the peak voltage across each switch. For the converters normally used with brushless d.c. motors the peak device voltage is nominally equal to the d.c. supply voltage, because each switch must block this voltage while the other one in the same phaseleg is conducting. Obviously there must be a margin for voltage spikes caused by stray inductance and reverse-recovery of diodes. While these effects are not small, they are parasitic and not fundamental to the operation of the motor. Therefore the d.c. supply voltage will be used in this comparison.

With respect to the peak current in each switch,

$$\text{peak kVA/kW} = 2q \times \hat{\imath}_s \times V_s$$

where $\hat{\imath}_s$ is the peak current in each switch, and the voltage conditions are unchanged.

In the sinewave motor the line currents are assumed to be sinewaves and each switch conducts a half sinewave for 180° and is then off for 180°. The r.m.s. switch current is therefore $1/\sqrt{2}$ times the r.m.s. line current, which will be assumed to be the same as the phase current (i.e. the motor is star-connected). The peak device current is equal to the peak phase current. The

# TORQUE PER AMPERE AND kVa/kW

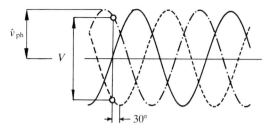

FIG. 5.9. Derivation of line voltage from d.c. supply voltage.

relationship between the d.c. supply voltage $V$ and the a.c. line voltage of the motor is derived from Fig. 5.9; the peak line-line voltage of the motor is equal to $V$. Thus

$$\hat{v}_{l-l} = V = \sqrt{3}\hat{v}_{ph} = \sqrt{6}V_{ph}.$$

We can now write:

$$\text{r.m.s. switch VA} = 6V\frac{I}{\sqrt{2}};$$

$$\text{peak switch VA} = 6V\hat{\imath};$$

$$\text{converter power output} = 3V_{ph}I = 3\frac{V}{\sqrt{6}}I.$$

From the appropriate ratios between these quantities, the r.m.s. switch kVA/kW is 3.5 and the peak kVA/kW is 6.9.

For the squarewave motor the corresponding equations are:

$$\text{r.m.s. switch VA} = 6Vi\sqrt{\frac{2}{3}};$$

$$\text{peak switch VA} = 6Vi;$$

$$\text{converter power output} = Vi.$$

From the appropriate ratios between these quantities, the r.m.s. switch kVA/kW is 4.9 and the peak kVA/kW is 6.0. The result is the same for the star- and delta-connected motors considered earlier.

Thus the squarewave motor has a slightly better utilization of the peak current capability of the converter switches. Although the sinewave motor appears from this analysis to have a much better utilization of their r.m.s. current capability, this advantage may be offset by the greater duty cycle in the sinewave motor, where three converter switches are conducting at any instant, not two. In the analysis it has been assumed that the ideal waveforms are achieved by pulse-width modulation at a sufficiently high frequency, and of course this incurs switching losses which tend to lessen the significance of the lower r.m.s. switch currents required by the sinewave motor. More detailed comparison requires the use of computer simulation.

### 5.6 Permanent magnets versus electromagnetic excitation

Because of the natural laws of electromagnetic scaling there is an 'excitation penalty' associated with small motors. As the geometrical size is decreased, the cross-sectional area available for copper conductors decreases with the square of the linear dimension, but the need for m.m.f. decreases only with the linear dimension, being primarily determined by the length of the airgap. As the motor size is further decreased, the airgap length reaches a minimum manufacturable value (typically of the order of 0.15 mm but often as much as 0.3 mm). Past this point the m.m.f. requirement decreases only slowly while the copper area continues to decrease with the squared linear dimension. The per-unit copper losses increase even faster, and the efficiency decreases rapidly. The loss-free excitation provided by permanent magnets therefore increases in relative value as the motor size is decreased.

In larger motors magnets can help to improve efficiency by eliminating the losses associated with electromagnetic field windings. But in larger motors the relative excitation penalty is small. At the same time the required volume of magnets with adequate properties increases with motor size to the point where PM excitation is just too expensive. It is therefore rare to find PM motors rated much larger than a few kilowatts.

Another reason why permanent magnets are not suitable for larger machines is that these machines often need a constant-power characteristic at higher speeds, or at least some field-weakening capability. Field-weakening is also desirable for minimizing iron losses especially at light loads, and while it can be applied to conventional synchronous and induction motors it is not possible to any useful extent with PM motors.

Sometimes there are operational or safety considerations that work against the PM motor. For example, in railway or transit-car traction, a PM motor with a short-circuit winding fault would have a large braking torque, possibly with a high ripple component and a definite risk of overheating which could demagnetize the magnets or even cause a fire. There is no simple way to protect against this type of fault. A conventional synchronous or d.c. machine can be de-excited under fault conditions, and the induction motor is self-protecting.

There is no hard-and-fast power level below which permanent-magnet excitation becomes advantageous, but it is possible to examine the excitation penalty in ways which indicate roughly where the breakpoint lies, and why. For a given level of excitation the choice can be made between magnets or copper windings operating at a current density $J$ (in the copper).

In the following analysis, several gross assumptions are made, and the equations derived should not be used for detailed design purposes but only for guidance and interpretation.

Using a permanent magnet, the fundamental flux per pole is $B_1 Dl/p$, and if rotor leakage is neglected this can be taken to be equal to the flux through the magnet in a surface-magnet motor, that is, $B_m A_m$. If the magnet is operated at a fraction $\gamma$ of its remanent flux-density, $B_m = \gamma B_r$ and hence

$$A_m = \frac{B_1 Dl}{p\gamma B_r}.$$

From Ampere's law we have

$$H_m l_m + H_g g = 0$$

from which

$$l_m = \frac{g B_g \mu_{rec}}{(1-\gamma) B_r}.$$

The required volume of magnet can now be determined as

$$V_m = 2p A_m l_m.$$

If $B_1 = k_1 B_g$, and if we assume that $l = D$, then

$$V_m = \frac{2 B_1^2 D^2 g \mu_{rec}}{k_1 B_r^2 \gamma (1-\gamma)}.$$

Now we can calculate the amount of copper needed to magnetize the airgap to the same level. Assuming a sinusoidal distribution of conductors,

$$F_g = g H_g = g \frac{B_g}{\mu_0} = \frac{g}{\mu_0} B_1 \cos p\theta.$$

But

$$F_g = \int_0^\theta i \frac{N}{2} \sin p\theta \, d\theta = \frac{iN}{2p} \cos p\theta.$$

Hence

$$\frac{g B_1}{\mu_0} = \frac{iN}{2p}$$

and the total ampere-conductors required are

$$iN \times 2p = 2piN.$$

If $A_c$ is the cross-section of the copper winding in the whole motor cross-section and $J$ is the current density,

$$A_c J = 2piN$$

and hence

$$A_c = \frac{4p^2 g B_1}{\mu_0 J}.$$

Note that the copper cross-section required is proportional to $p^2$, unlike the magnet volume which is independent of the number of poles. The volume of copper can be estimated by assuming a mean length of conductor equal to twice the stack length (to allow for end-turns); thus

$$V_c = \frac{8p^2 g D B_1}{\mu_0 J}.$$

The relative volumes of magnet and copper required can be compared by taking the ratio

$$\frac{V_m}{V_c} = \frac{\mu_0 \mu_{rec} J B_1 D}{4 k_1 B_r^2 p^2 \gamma (1-\gamma)}$$

where $D$ is the rotor diameter (assumed equal to half the stator diameter).

Consider two four-pole motors with $B_1 = 0.7$ T and $k_1 = 1.1$. The PM motor has rare-earth magnets with $B_r = 0.8$ T and $\mu_{rec} = 1.05$. The electrically-excited motor has $J = 4$ A/mm$^2$, giving

$$\frac{V_m}{V_c} = \frac{D}{488}$$

where $D$ is measured in mm. This means that for motors less than about 500 mm in rotor diameter the magnet volume is less than the volume of copper needed for excitation by a separate field winding. Unfortunately the cost per unit volume of high-energy magnets at this level is of the order of 25 times that of copper. For the magnet cost to be less than the cost of the copper in a separate field winding, the rotor diameter must therefore be less than 500/25, i.e. only 20 mm, giving a stator diameter of about 40 mm. The technical potential of high-energy magnets is thus offset by their high cost in all but the smallest motors. In very small motors a smaller value should be used for the current density $J$; with $J = 2.5$ A/mm$^2$ the stator diameter for equal cost would be increased from 40 mm to 64 mm. In general, high-energy magnets can only be justified where there is a special premium on efficiency or compactness. Of course this argument is simplistic, ignoring factors such as process and manufacturing costs and many others, but it provides a basic physical understanding of the application potential of magnets, and the effects of scale.

Motors magnetized with ceramic magnets must settle for a lower airgap flux-density. Using values of $J = 4$ A/mm$^2$; $B_1 = 0.3$ T; $\mu_{rec} = 1$; $B_r = 0.35$ T, the result is

$$\frac{V_m}{V_c} = \frac{D}{230}.$$

For motors of less than 460 mm stator diameter the magnet volume indicated is less than the volume of copper in a separate field winding. Ceramic magnets are much less expensive than high-energy magnets, the cost per unit volume being of the order of 0.6 times that of copper, so that the magnet cost will be less than the cost of field copper in motors of diameter less than $460 \times 0.6$, i.e. 275 mm. In practice PM motors as large as this are relatively uncommon. With ferrite, the flux-density is too low; with rare-earth magnets the cost is too high.

If running costs are taken into account, the comparison between PM and electrically excited motors changes significantly. With the present cost of raw materials and present kWh tariffs, the kWh cost of electrical excitation would outstrip the raw-material cost of the copper in just a few months, assuming the motor runs at full excitation 24 hours per day. Even when all the manufacturing costs are added up, the PM motor should eventually pay for itself in this way.

## 5.7 Slotless motors

Recently the availability of very high energy rare-earth and neodymium–iron–boron magnets has re-awakened interest in the slotless motor, in which the stator teeth are removed and the resulting space is partially filled with additional copper. At least one such motor is manufactured commercially. The slotless construction permits an increase in rotor diameter within the same frame size, or alternatively an increase in electric loading without a corresponding increase in current density. The magnetic flux-density at the stator winding is inevitably lessened, but the effect is not so drastic as might be expected. For a motor with an iron stator yoke and an iron rotor body the magnetic field and its harmonic components can be calculated by the methods described by Hughes and Miller (1977). Considering the fundamental radial component of $B$, the value is greatest at the rotor surface (radius $r$) and falls off with increasing radius to its smallest value just inside the stator yoke (radius $R$). The ratio between the values of the fundamental radial component at these two radii is given by:

$$b = \frac{2(r/R)^{p+1}}{[1 + (r/R)^{2p}]}.$$

Consider a rotor body of 40 mm diameter with a high-energy magnet of remanent flux-density 1.2 T and thickness 5 mm. If the radial thickness of the stator winding is 5 mm (including the airgap), then for a four-pole magnet $b = 0.78$. The magnet flux-density will be about half the remanent flux-density with these proportions, so that the radial flux-density in the stator winding varies from about 0.6 T near the bore to 0.47 T just inside the stator yoke, giving a mean value of fundamental flux-density of about 0.53 T. The electric loading may be increased relative to that of a slotted stator, because of the additional space available for copper; but the increase may not be much because the close thermal contact with the teeth is lost, and the cooling of the slotless winding by conduction to the stator steel may not be as effective. Taking these factors into account, the power density should be roughly the same as that of the conventional motor, and possibly a little higher, since the stator tooth iron losses are eliminated. This machine may well accept less expensive grades of lamination steel because of the absence of slotting and the relatively low flux-density in the stator yoke. The reactance is also lessened by the elimination of slot leakage effects, and the risk of demagnetization is decreased.

In this type of motor the maximum useable magnet energy is obviously higher than in a conventional slotted motor; indeed the concept would not be viable at all without magnets of high remanence and coercivity.

Once the stator teeth are removed, the conductors are no longer constrained to lie parallel to the axis. They may be skewed by a small amount to reduce torque ripple (which is already reduced by the elimination of cogging effects against the stator teeth). A further possibility is a completely helical winding such as that proposed for superconducting a.c. generators (Ross 1971), or as used in very small PM commutator motors. Because the helical winding has no end-turns its utilization of copper is higher than the severe skew might suggest, and it might permit the design of a very compact motor

## 5.8 Ripple torque in sinewave motors

If the stator current waveform is not a pure sinewave, and if the conductor distribution is not purely sinusoidal around the airgap, then the time harmonics of the current can interact with the space-harmonics of the conductor distribution to produce constant space-harmonic components of the ampere-conductor distribution that have the same pole-pitch as the corresponding harmonics of the rotor flux distribution; and if both rotate at synchronous speed there will be a contribution to the average torque in addition to that produced by the fundamental components. These harmonic torques may add up to a few percent of the fundamental component. Unfortunately associated with them are space-harmonic components of the ampere-conductor distribution that rotate at non-synchronous speeds and produce torque ripple. A thorough analysis of these harmonic torques was published by Bolton and Ashen (1984). They calculated average and ripple torques for several motor configurations and different phase current waveforms. With a 160° trapezoidal magnet flux distribution and a three phase, full-pitched winding with one slot per phasebelt and no skew, they calculated a peak–peak torque ripple of 14 per cent with pure sinewave currents, rising to 37 per cent with 180° squarewave phase currents, relative to the average torque of a machine with purely sinusoidal windings and magnet flux. Relative to the average torque produced in each of the two cases, the calculated torque ripple is 5.2 per cent with sinewave currents and 19.4 per cent with 180° square-wave currents. With more than one slot per phasebelt and skewed slots, the torque ripple would be less. These figures are quoted to indicate the kind of torque ripple components that might be expected; in practice it is extremely difficult to calculate torque ripple accurately at these levels, because of the interaction of many effects, including cogging, magnetic fringing, saturation, and other parasitic effects. Bolton and Ashen's analysis also highlighted the very smooth torque production of the nine-phase motor.

The lowest characteristic frequency of ripple torque due to winding and flux-distribution harmonics is shown by Bolton and Ashen to be $2q$ cycles per pole-pair of rotor rotation for a machine with an odd number of phases, $q$. When $q$ is even, the lowest characteristic frequency is $q$ cycles per pole-pair of rotor rotation.

## Problems for Chapter 5

1. A three-phase, four-pole brushless PM Motor has 36 stator slots. Each phase winding is made up of three coils per pole with 20 turns per coil. The coil span is seven slots. If the fundamental component of magnet flux is 1.8 mWb, calculate the open-circuit phase e.m.f. $E_q$ at 3000 r.p.m.

2. The stator bore diameter of the motor of Problem 1 is 100 mm and its axial length is 20 mm. The airgap is 1 mm, including an allowance for slotting. The magnet has a radial thickness of 10 mm and a recoil permeability of 1.01. Calculate the airgap synchronous reactance at 100 Hz.

3. In the motor of Problem 1, what is the torque at 3000 r.p.m. if the phase current is 4.0 A and $\gamma = 0$?

4. In the motor of Problem 1, the phase resistance is 3.7 $\Omega$ and the leakage reactance is 5 mH. With the phase current of 4.0 A at $\gamma = 0$ at 3000 r.p.m., calculate the terminal voltage and the power-factor angle.

5. Repeat Problem 4 with $I = 4.0$ A and $\gamma = -15°$.

6. A brushless PM sinewave motor has an open-circuit voltage of 173 V at its corner-point speed of 3000 r.p.m. It is supplied from a p.w.m. converter whose maximum voltage is 200 V r.m.s. Neglecting resistance and all other losses, estimate the maximum speed at which maximum current can be supplied to the motor.

# 6 Alternating-current drives with PM and synchronous-reluctance hybrid motors

In Chapter 1 (Fig. 1.4) the synchronous PM a.c. motor was derived from the conventional wound-field version. If a cage winding is included in the rotor, or if the rotor is fabricated from solid steel, then the motor can start, synchronize, and run from an a.c. sinewave supply without an electronic converter. It operates as a synchronous reluctance motor if the magnets are left out or if they are demagnetized.

The motors in this class are hybrid PM/synchronous reluctance motors. They have several useful properties, including:

(1) combined reluctance and magnet alignment torque;
(2) field-weakening capability;
(3) under-excited operation for most load conditions;
(4) high inductance;
(5) high speed capability; and
(6) high temperature capability.

The synchronous reluctance motor is completely free of magnets and their operational problems. It is inexpensive to make, and can operate at extremely high speeds and at higher temperatures than PM motors. However, its power factor and efficiency are not as high as those of a PM motor, and the converter kVA requirement is higher. The synchronous reluctance motor offers many of the advantages of the switched reluctance motor discussed in Chapter 7, but with the two added advantages that it can operate from essentially standard p.w.m. a.c. inverters and has lower torque ripple. It can also be built with a standard induction-motor stator and winding.

## 6.1 Rotors

Figures 6.1 to 6.4 show a variety of rotors and components, including both solid-steel and laminated versions. The motors in Figs 6.1 and 6.2 are both cageless, Fig 6.3 shows a line-start motor with slots for a cast aluminium starting cage, and Fig. 6.4 an unusual brushless synchronous motor that can be excited either by magnets or field windings located in the rotor extensions.

Figure 6.1 shows the two basic configurations of 'interior magnet' motors although several others are used. In Fig. 6.1(a) (Jahns et al., 1986) the magnets are alternately poled and radially magnetized, but because the magnet pole area is smaller than the pole area at the rotor surface, the airgap flux-density

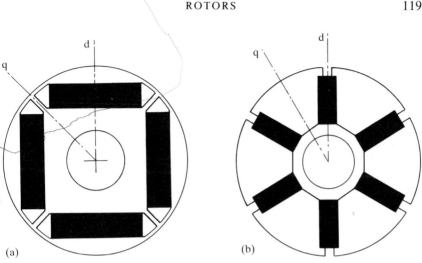

FIG. 6.1. Cross-sections of cageless PM/synchronous reluctance hybrid motors. (a) Four-pole radially magnetized. Without the magnets, this motor is a pure synchronous reluctance motor. (b) Six-pole circumferentially magnetized. This arrangement is known as a 'flux-concentrating' design because the magnet pole area exceeds the pole area at the airgap, producing an airgap flux-density higher than that in the magnet. The stators of these motors are essentially the same as for the polyphase induction motor.

on open-circuit is less than the flux-density in the magnet; this design is therefore essentially 'underexcited' and relies on the addition of a magnetizing component of armature current to produce the total airgap flux. Because there is an appreciable permeance to q-axis flux and a low permeance to d-axis armature-reaction flux, this machine has considerable reluctance torque and field-weakening capability, providing a constant-power characteristic at high speeds. The machine was developed in this form by Jahns, Kliman, and Neumann (1986). It is a true a.c. motor and cannot be operated as a square-wave brushless motor. This follows from the combination of magnet-alignment and reluctance torques. Both of these torque components can be kept constant only with a fixed load angle, and this requires a rotating field of the type that is possible only with sine-distributed windings and sinusoidal phase currents.

The magnets in Fig. 6.1(b) are circumferentially magnetized and alternately poled, so that two magnets communicate flux to each pole piece; the sum of two magnet pole areas exceeds the pole area at the rotor surface, producing an open-circuit airgap flux-density greater than that in the magnet. For this reason the term 'flux-concentrating' or 'flux-focusing' is sometimes used to describe it. A nonmagnetic shaft or spacer is needed to prevent the magnets from being short-circuited at their inner edges, and with this the permeance to q-axis flux is very low. This machine therefore has little reluctance torque. Correspondingly the field-weakening capability is limited and so is the speed

120 ALTERNATING CURRENT DRIVES

FIG. 6.2. Components of PM/reluctance hybrid motor showing the wound stator, partially assembled rotors, laminations, and permanent magnets. The 'smartpower' integrated circuit at bottom right contains all the power electronics needed to run this motor at 40 W from a 60 V d.c. supply. Courtesy Scottish Power Electronics and Electric Drives, Glasgow University; General Electric; and the Science and Engineering Research Council.

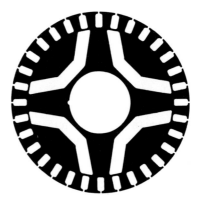

FIG. 6.3. Line-start synchronous reluctance motor rotor. Courtesy Professor P. J Lawrenson, Switched Reluctance Drives Ltd.

FIG. 6.4. Hybrid PM/electrically excited motor with salient poles. This brushless synchronous machine can be excited either by magnets or field windings projecting into the rotor extensions. Courtesy Magnetics Research International, Iowa, USA.

range at constant power. It can be operated very well as a brushless d.c. motor with square-wave excitation, and in this form it is not a true hybrid motor. Apart from the calculation of its magnetic circuit, its features are similar to those of the motors in Chapters 4 and 5 (see Binns 1984; Binns and Kurdali 1979).

In this chapter the emphasis is on the synchronous reluctance and hybrid characteristics. Figure 6.5(b) shows, schematically, the open-circuit flux in the PM version, and Fig. 6.7(b) shows the no-load flux in the pure reluctance motor when there is no torque; all the flux is q-axis flux at no-load, and there is no d-axis flux. The d-axis in the reluctance motor is taken to be the same as in the PM motor. This is contrary to the convention adopted in some early reluctance-motor analysis, in which the d-axis and q-axis were interchanged. With the convention used here it is usually the case that

$$X_d < X_q$$

which is the opposite of the situation in wound-field synchronous machines.

The use of a one-piece lamination requires compromises in the design. The section linking the pole pieces must be wide enough to support them and the magnets against the centrifugal load, but narrow enough to limit its participation in the magnetic circuit, which is undesirable in both the reluctance and PM versions. If it is too thin, the lamination is flimsy and easily damaged. A cast cage winding, or even a fabricated one, may help to relieve some of the mechanical constraints on the lamination design, but a cage winding may not be desired. The rotor is amenable to fabrication from solid steel parts with no connecting links between the poles; instead, the pole pieces are held on by end-caps at each end of the rotor. (This construction is not shown.) The rotors in Fig. 6.2 were machined by wire-EDM (electro-discharge machining). This technique is too expensive for mass-production but very convenient for small quantities.

The motor in Fig. 6.4 is unusual in that it can be excited either by permanent magnets or by field windings accommodated in the rotor extensions. The magnets can be integral with the rotor, and rotate with it; or they can be stationary, in which case the flux must cross the airgap twice. In the electromagnetically excited version the field windings are stationary, so there are no brushes or sliprings.

## 6.2 A.c. windings and inductances

The following analysis treats the PM and reluctance motors together; the reluctance motor can then be studied as a special case with the magnets removed. The analysis follows the same course as for the non-salient-pole PM motor, but with the added complications arising from the rotor geometry and its two-axis symmetry.

Figures 6.5–6.7 show the model for analysis with a single sine-distributed winding aligned with the direct axis of the rotor. As in Chapter 5, we will later substitute the effective sine-distributed turns in series per phase; but to determine the interaction of the fundamental flux and ampere-conductor distributions, the pure sinewave model is analysed first. The model has four poles, but the equations developed are in general applicable to machines with any number of poles, except for slight anomalies in the two-pole case.

### 6.2.1 Open-circuit e.m.f.

The stator and rotor steel is assumed to be infinitely permeable everywhere except in the link sections between the pole pieces. This permits the surfaces to be represented by equipotentials. The q-axes are also equipotentials, by symmetry. Figure 6.5(a) therefore defines a model in which there are only two boundary potentials, $u_0$ and $u_1$. The potential function here is magnetic potential, and its units are amperes (or ampere-turns). The link sections are assumed to saturate at a flux-density $B_s$. On open-circuit the flux through them is leakage flux that follows a closed path through the rotor and does not cross the airgap. The equipotentials are assumed not to be distorted by this leakage flux.

We can arbitrarily assign one of the two potentials to be zero, because the fluxes of interest depend on potential differences, not on absolute potential

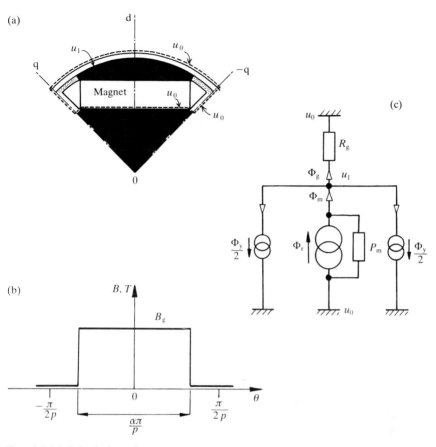

FIG. 6.5 (a) Calculation of open-circuit flux-density due to the magnets, showing the assumed magnetic potential boundaries. (b) Distribution of magnet flux in the airgap across one pole. (c) Magnet equivalent circuit.

values. Since the rotor shaft and the stator are both common to all the poles, it seems sensible to assign

$$u_0 = 0.$$

However, in some of the equations we will retain $u_0$ when we want to make it clear that we are considering a potential difference.

If the airgap is small and if fringing is neglected, the radial flux-density in the gap is

$$B_g = \frac{\mu_0}{g'}(u_1 - u_0) = \frac{\mu_0}{g'} u_1 \text{ T}.$$

This gives rise to a rectangular distribution of flux across the pole, as shown in Fig. 6.5(b).

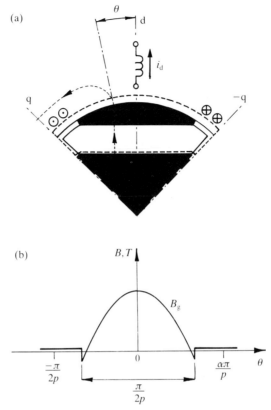

FIG. 6.6. (a) Calculation of d-axis synchronous reactance, showing the assumed magnetic potential boundaries. (b) Distribution of d-axis flux excited by sine-distributed stator winding.

In terms of the reluctance of the airgap,

$$\Phi_g = u_1 P_g$$

where the airgap permeance is given by

$$P_g = \frac{1}{R_g} = \frac{\mu_0 A_g}{g'}.$$

If the pole-arc/pole-pitch ratio is $\alpha$, the stator bore radius is $r_1$, and the stack length is $l$,

$$A_g = \alpha \frac{\pi}{p} r_1 l.$$

On the underside of the pole-piece the magnet can be represented as a magnetic equivalent circuit as shown in Fig. 6.5(c). The link sections are

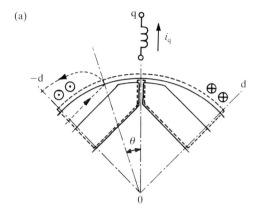

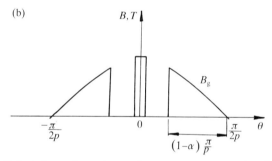

FIG. 6.7. (a) Calculation of q-axis synchronous reactance, showing the assumed magnetic potential boundaries. (b) Distribution of q-axis flux excited by sine-distributed stator winding.

included in the equivalent circuit as flux sources, one on each side, each carrying the fixed leakage flux

$$\frac{1}{2}\Phi_y = B_s yl.$$

The magnetic p.d. across the magnet is the same as that across the airgap and therefore

$$u_1 = \frac{\Phi_r - \Phi_y}{P_m + P_g}$$

where $\Phi_r$ is the magnet remanent flux:

$$\Phi_r = B_r A_m$$

and $A_m$ is the pole area of the magnet:

$$A_m = w_m l.$$

The effective permeance $P_m$ assigned to the magnet includes a component due to additional rotor leakage flux which passes between adjacent poles without crossing the airgap. This leakage flux flows in a circumferential direction through the air under the link sections, and is in parallel with the flux $\Phi_y$. Rather than model it by a separate permeance, it is incorporated in a modified magnet permeance calculated as though the magnet pole width were increased to

$$w'_m = \mathrm{av}(w_m, w_m + h) = w_m + \frac{h}{2}.$$

This gives

$$A'_m = w'_m l$$

and

$$P_m = \frac{\mu_{\mathrm{rec}} \mu_0 A'_m}{l_m}.$$

Hence the airgap flux is given by

$$\Phi_g = u_1 P_g = \frac{\Phi_r - \Phi_y}{1 + P_m R_g} = B_g A_g.$$

The pole-piece can be regarded as a 'potential island'. Its magnetic potential is determined by the properties of the magnet, by the cross-sectional geometry, and by the flux leaking through the link sections as represented in Fig. 6.5. This is different from the situation in conventional machines where there is only one contiguous region of highly permeable material.

The model is a 'per-pole' model and is the same for every pole in the machine. The magnetic potential of the adjacent pole-pieces is $-u_1$, by symmetry, so that the potential difference between two adjacent pole-pieces is $2u_1$.

The rectangular flux-distribution in the airgap is identical to the ideal distribution calculated for the surface-magnet motor in Chapter 4. This suggests that the motor could be driven as a square-wave motor as in Section 4.6. This would be true if there were no reluctance torque. We have already seen that, for the reluctance torque to be zero, $X_d$ and $X_q$ must be equal. In the structure of Fig. 6.1 they are in general unequal. In theory there may be some cross-section that would make them equal, but in practice it would be difficult to keep them equal if the steel saturates to any degree. Moreover, if the motor of Fig. 6.1(a) is to be driven as a squarewave motor, it is clear that the open-circuit airgap flux-density will be lower than if the magnets were on the rotor surface, because the magnet pole area is appreciably smaller than the pole face area and there is an inverse concentration of flux-density. When these factors are considered in depth, it becomes clear that the hybrid motor is not a good square-wave motor, and it is better to design it for sinewave operation so that combination of magnet alignment torque and reluctance torque can be optimized.

The fundamental open-circuit flux per pole can be determined by Fourier analysis of the waveform in Fig. 6.5(b):

$$\Phi_{M1} = \frac{B_{M1} Dl}{p}.$$

The amplitude of the fundamental component of the airgap flux due to the magnet acting alone is

$$B_{M1} = k_1 B_g$$

where

$$k_1 = \frac{4}{\pi} \sin \frac{\alpha\pi}{2}.$$

A further result that is sometimes useful is the relationship between the fundamental flux per pole and the actual flux per pole. With rectangular distribution of flux the result is

$$\Phi_{M1} = \Phi_g \frac{8}{\pi^2 \alpha} \sin \frac{\alpha\pi}{2}.$$

This formula can, of course, be used for any rectangular distribution of airgap flux, not just the magnet flux. It is a familiar result from the theory of synchronous machines (see Wieseman 1927).

Now the open-circuit e.m.f. per phase can be determined exactly as in Chapter 5. For a practical winding with $N_{ph}$ series turns per phase and a winding factor $k_{w1}$, the result is

$$E_q = \frac{2\pi}{\sqrt{2}} (k_{w1} N_{ph}) \Phi_{M1} f.$$

This equation can also be expressed in the form

$$\tilde{E}_{ph} = jE_q = j\omega \tilde{\Psi}_{M1}$$

where

$$\tilde{\Psi}_{M1} = \frac{1}{\sqrt{2}} \tilde{\psi}_{M1} = \frac{1}{\sqrt{2}} k_{w1} N_{ph} \Phi_{M1}.$$

The subscript 'q' has exactly the same meaning as in Chapter 5, that the phasor e.m.f. lies on the q-axis of the phasor diagram.

### 6.2.2 Synchronous reactance (d-axis)

The calculation of d-axis synchronous reactance follows the same procedure as in Chapter 5. The model is shown in Fig. 6.6, which shows a typical assumed flux path along which Ampere's law is applied. The magnet is assumed to be present and represented by its recoil permeability. The armature reaction flux is assumed to be superimposed on the flux produced by the magnet. Note the positions and directions of the currents, representing the ampere-turns produced by $N_p$ turns per pole distributed in two slots per pole per phase.

The narrow link sections between the poles are assumed to be permanently saturated by the magnet in the 'recirculating' direction defined in the open-circuit condition. Negative d-axis current in the stator tends to reinforce this flux, whereas positive d-axis current tends to oppose it. The sign of $\Phi_y$ could therefore reverse when the d-axis current reaches some positive value.

In normal operation the d-axis current is negative, i.e. demagnetizing. It may be positive at light loads, but in a well designed system the terminal voltage should then be low enough to keep the d-axis current fairly small. Therefore it is assumed that the magnet maintains the links in saturation over all normal operating conditions, and that they have no effect on the superposed armature-reaction flux.

In the pure reluctance motor with no magnets, the links are assumed to be saturated permanently by q-axis armature current and are omitted from the calculation of d-axis reactance. In small motors this may not actually be the case, but in such small machines the airgap component of $X_d$ may be appreciably smaller than the leakage inductance, in which case the error arising from the neglect of the links is acceptably small.

With a sine-distribution of ampere-conductors whose magnetic axis is aligned with the d-axis,

$$\frac{N_s}{2} \sin p\theta$$

the m.m.f. integral $\int H \, dl$ for flux-lines that cross the airgap via the pole face is given by

$$\frac{N_s i}{2p} \cos p\theta$$

where

$$-\frac{\alpha\pi}{2p} < \theta < \frac{\alpha\pi}{2p}.$$

This expression equals one-half the ampere-conductors enclosed within a closed flux-line that crosses the airgap at the angle $\theta$. The other half of the

## A.C. WINDINGS AND INDUCTANCES

enclosed ampere-conductors can be thought of as forcing the flux line across the airgap via the adjacent pole. Thus the equations developed here are all on a 'per-pole' basis. If all the poles are in series, $N_s$ is the number of turns in series per phase, and $N_s/2p (= N_p)$ is the number of turns per pole, as in Chapter 5. Flux entering the sides of the pole is classed as fringing flux and is ignored at this stage.

The dotted line drawn across the rotor and along the q-axes in Fig. 6.6(a) is an equipotential $u_0$ and, as before, this potential may be assigned to zero with no loss of generality, since it is common between adjacent poles. The pole-piece is at a uniform magnetic potential $u'_1$, as yet unknown. The stator bore is no longer at the uniform potential $u_0$ but has a potential $u_s$ that varies with $\theta$, its gradient sustained by the ampere-conductor distribution; evidently

$$u_s(\theta) = \frac{N_s i}{2p} \cos p\theta = \hat{u}_s \cos p\theta$$

where

$$\hat{u}_s = \frac{N_s i}{2p}.$$

The flux-density across the airgap is therefore

$$B_g(\theta) = \frac{\mu_0}{g'}(u_s(\theta) - u'_1) = \frac{\mu_0}{g'}(\hat{u}_s \cos p\theta - u'_1).$$

This distribution is plotted in Fig. 6.6(b). The offset or 'd.c. component' is caused by the induced pole-piece potential $u'_1$. Notice that with a wide enough pole and sufficient negative (demagnetizing) d-axis current it is possible for the radial flux density near the edges of the pole to be reversed. In practice fringing flux will smooth out the sharper changes of magnetic potential around the pole edges.

The flux per pole is the integral over the pole face:

$$\Phi_p = \int B_g r_1 \, d\theta l$$
$$= \frac{\mu_0 D l}{p g'}\left(\hat{u}_s \sin \frac{\alpha\pi}{2} - u'_1 \frac{\alpha\pi}{2}\right).$$

This equation can be rewritten as

$$\Phi_p = \frac{\hat{u}_s k_{\alpha d} - u'_1}{R_g}$$

where

$$k_{\alpha d} = \frac{\sin(\alpha\pi/2)}{\alpha\pi/2}.$$

The undetermined potential $u'_1$ is determined by applying Gauss's law to the pole-piece; that is, by equating the flux entering via the pole-face ($\Phi_p$) to the flux leaving via the magnet and the parallel rotor leakage permeances $P_r/2$ at each side. Thus

$$(u'_1 - u_0)P_m \pm \Phi_y = \Phi_p$$

from which

$$u'_1 = \frac{k_{\alpha d}}{1 + P_m R_g} \hat{u}_s.$$

The fundamental flux per pole is determined by Fourier analysis of the airgap flux distribution: making use of symmetry the result is

$$\hat{B}_{a1} = \frac{4}{\pi} \int_0^{\pi/2p} B_g(\theta) \, d(\theta)$$

$$= \frac{\mu_0}{g'} [k_{1ad}\hat{u}_s - k_1 u'_1]$$

$$= \frac{\mu_0}{g''} \hat{u}_s$$

where

$$g''_d = \frac{g'}{k_{1ad} - \dfrac{k_1 k_{\alpha d}}{1 + P_m R_g}}$$

is the 'effective airgap in the d-axis' and

$$k_{1ad} = \alpha + \frac{\sin \alpha \pi}{\pi}.$$

The fundamental armature-reaction flux per pole can be derived exactly as in Section 5.1, and indeed the whole of the remaining derivation of the synchronous reactance is the same. If all the poles are in series we can substitute the effective number of sine-distributed turns per phase and the formula for the synchronous reactance is exactly the same as in Section 5.2; only the value of $g''$ is different. Thus with

$$N_s = \frac{4}{\pi} k_{w1} N_{ph}$$

we have

$$X_d = \frac{6\mu_0 D l f}{p^2 g''_d} (k_{w1} N_{ph})^2 + X_\sigma.$$

For a two-phase machine the airgap reactance has $\frac{2}{3}$ the value calculated by this formula. Note that the effective airgap in the above formula does not account for the flux in the link sections.

## 6.2.3 Synchronous reactance (q-axis)

When the stator ampere-conductor distribution is aligned with the q-axis the flux produced by it does not pass through the magnet; see Fig. 6.7(a). The magnetic potential distribution along the stator bore is the same as in the d-axis case, but displaced 90° ahead. The rotor steel surfaces lie on equipotentials that are completed by sections of the d-axes on either side. The web and the two link sections are modelled by one of two different methods, depending whether they are saturated or not. If they are unsaturated, they can be considered as a highly permeable flux guide of width $\Omega\pi$ electrical radians and magnetic potential $u_0$ at the airgap. This gives rise to an ideal flux distribution in the airgap shown in Fig. 6.7(b). The fundamental component of this is $k_{1aq}$ times the peak value on the q-axis, where

$$k_{1aq} = \alpha + \Omega + \frac{\sin \Omega\pi - \sin \alpha\pi}{\pi}.$$

Hence

$$X_q = \frac{6\mu_0 Dlf}{p^2 g_q''}(k_{w1}N_{ph})^2 + X_\sigma$$

where

$$g_q'' = \frac{g'}{k_{1aq}}$$

is the 'effective airgap in the q-axis'. For a two-phase machine the airgap reactance has $\frac{2}{3}$ the value calculated by this formula. In the example (Table 6.1) the value of $k_{1aq}$ without the web would be 0.535, so the web adds 37 per cent to the airgap component of $X_q$. The effective value of $\Omega$ was estimated as 0.1 to allow for fringing effects and the contribution of the links.

The effective airgaps $g_d''$ and $g_q''$ are a convenient way of incorporating the pole and magnet geometry into the formulas for synchronous reactance, because they put the PM machine on the same basis as the conventional synchronous machine. They also unify all the synchronous PM machines by including the surface-magnet machine as a special case of the more general salient-pole machine considered here.

The airgap reactance ratio for the machine in Figs 6.1 and 6.2 is

$$\frac{X_q}{X_d} = \frac{1.818}{0.529} = 3.44$$

before the leakage reactances are added. The higher this ratio, the greater the potential reluctance torque. It is difficult to obtain practical values much higher than that of the example motor.

If the web and link sections are saturated the fluxes in them are

$$\phi_y = B_s yl \text{ and } \Phi_w = B_s wl.$$

**Table 6.1.** Parameters of the motor of Fig. 6.2

| Parameter | Symbol | Value | Unit |
|---|---|---|---|
| Operating frequency | $f$ | 100 | Hz |
| Number of phases | $q$ | 2 | |
| Pole pairs | $p$ | 2 | |
| Pole arc | | 68.0 | degrees |
| Pole arc/pole pitch | $\alpha$ | 0.756 | |
| Stator bore radius | $r_1$ | 20.7 | mm |
| Rotor radius | | 20.25 | mm |
| Stack length | $l$ | 50.8 | mm |
| Airgap | $g$ | 0.45 | mm |
| Carter coefficient | $K_c$ | 1.36 | |
| Effective airgap | $g'$ | 0.612 | mm |
| Magnet remanent flux-density | $B_r$ | 1.1 | T |
| Recoil permeability | $\mu_{rec}$ | 1.05 | |
| Magnet width | $w_m$ | 20.0 | mm |
| Magnet length | $l_m$ | 5.4 | mm |
| Web width | $w$ | 1.0 | mm |
| Link width | $y$ | 0.5 | mm |
| Number of series turns per phase | $N_{ph}$ | 96 | |
| Fund. winding factor | $k_{w1}$ | 0.924 | |
| Assumed saturation flux-density of steel | $B_s$ | 1.8 | T |
| Parameters at 100 Hz (3000 r.p.m.) | $E_q$ | 35.8 | V |
| | $E_{dw}$ | 0 | V |
| | $R$ (at 20°C) | 0.464 | $\Omega$ |
| | $X_\sigma$ | 0.65 | $\Omega$ |
| | $X_d$ | 1.18 | $\Omega$ |
| | $X_q$ | 2.47 | $\Omega$ |

The lower-case $\phi$ denotes the flux in only one link section. For the complete pole there are two link sections to take into account. In the pure reluctance motor, if all the current is in the q-axis, the link fluxes are oriented symmetrically about the q-axis (see Fig. 6.7(a)), and for the complete pole

$$\Phi_y = 2\phi_y.$$

In the permanent-magnet motor, link fluxes are established by the magnet on open-circuit; when q-axis current flows, it tends to reinforce the flux in the link connecting to the leading edge of the lagging pole, and oppose it in the link connecting to the lagging edge of the leading pole. In this case it is assumed that the magnet m.m.f. maintains the link fluxes, such that one is always directed towards the q-axis and the other one away from it, and there is no contribution to the q-axis flux-linkage of the stator winding.

Unlike the link fluxes the web flux is, ideally, unaffected by the magnet or the d-axis stator current, and can be added to the q-axis flux per pole. Figure 6.7(b) shows the approximate distribution of these fluxes in crossing

## A.C. WINDINGS AND INDUCTANCES

the airgap, in either the pure reluctance motor or the PM hybrid motor. The width of the field of flux in Fig. 6.7(b) is arbitrarily made equal to $2(w+2y)$ in the reluctance motor and $2w$ in the PM motor, the factor 2 representing an arbitrary allowance for the diffusion or fringing effect as the flux crosses the airgap. The contribution to the flux per pole is

$$\Phi_y + \Phi_w$$

in the pure reluctance motor and $\Phi_w$ in the PM motor. This flux has an approximately rectangular distribution in the airgap and by an earlier result in this section the fundamental flux per pole is given by

$$\Phi_{w1} = \frac{8}{\pi^2 \xi} \sin \frac{\xi \pi}{2}$$

where $\xi$ is the 'pole-arc' of the web flux field in the airgap, normalized to the pole-pitch. If $\xi$ is small the approximation $\sin x = x$ can be used to simplify this to

$$\Phi_{w1} = \frac{4}{\pi} (\Phi_y + \Phi_w).$$

Under saturated conditions the web flux (or combined web and link flux) is not proportional to the q-axis stator current, and it is not convenient to try to incorporate it in a modified $X_q$. Instead, it can be treated as a fixed q-axis flux-linkage component that generates a voltage in the d-axis of the phasor diagram, in much the same way that the magnet flux generates $E_q$. Thus

$$E_{dw} = \frac{2\pi}{\sqrt{2}} (k_{w1} N_{ph}) \Phi_{w1} f.$$

This result can be meaningful, of course, only if there is enough q-axis m.m.f. to saturate the web and the links. The necessary q-axis phase current to do this can be estimated using Ampere's law. If $N_p$ is the number of turns per pole,

$$B_g = \frac{\mu_0 N_p i_q}{g'}.$$

To saturate the web $B_g$ must exceed 1.8 T and in the example motor this requires 18.3 A of q-axis current—far more than the thermal capability of the stator winding. In this machine it is reasonable to assume that the web and links remain unsaturated other than by magnet flux.

For larger motors the above results can be expressed in the following form for use in the phasor diagram. The peak phase flux-linkage due to the web flux is

$$\hat{\psi}_{wq1} = k_{w1} N_{ph} \Phi_{w1} \text{ V s.}$$

The subscript 'q' means that in any phase this flux-linkage peaks when the axis of that phase is aligned with the q-axis. In the phasor diagram it means that the flux-linkage is due to the q-axis component of armature current. Then the e.m.f. is given by

$$E_{dw} = -\frac{1}{\sqrt{2}} \omega \hat{\psi}_{wq1}.$$

Since the e.m.f. must lead the flux-linkage which induces it by 90° in the phasor diagram, it appears negative because the q-axis leads the d-axis. The second subscript 'w' denotes that the e.m.f. is caused by the web flux.

### 6.2.4 Magnet flux-density and operating point

The magnetic potential difference across the magnet is

$$F_m = H_m l_m = -(u_1 + u'_1)$$

where $u_1$ is the potential of the pole-piece induced by the magnet, and $u'_1$ is the potential induced by armature current in the d-axis. The condition of concern is when $I_d$ is negative or demagnetizing, in which case $u_1$ and $u'_1$ are additive and force the operating point of the magnet down the demagnetization curve. The calculation of this operating point should assume the highest working temperature, and the above equation is then the basis for determining the minimum magnet thickness that can be used without the operating point descending below the knee of the demagnetization curve. If this happens there will be an irreversible loss of magnetization.

The potentials $u_1$ and $u'_1$ are given by

$$u_1 = \frac{\Phi_r - \Phi_y}{P_m + P_g}$$

and

$$u'_1 = \frac{k_{\alpha d}}{1 + P_m R_g} \left( \frac{1}{2p} \frac{4}{\pi} k_{w1} N_{ph} \right) \frac{3}{2} I_d \sqrt{2}.$$

For a two-phase motor the $\frac{3}{2}$ factor is omitted. The term in large brackets is the effective sine-distributed turns per pole, and it is assumed that $I_d$ flows through all of them in series.

The value of $I_d$ to be used depends on what is considered to be a 'worst case'. For electronically controlled motors it is reasonable to assume a terminal short-circuit on all phases with the motor at its maximum speed and maximum temperature. If resistance is neglected, then from the phasor diagram the short-circuit current is

$$I_{d(s.c.)} = \frac{E_q}{X_d}.$$

In small motors the resistance increases the internal short-circuit impedance and shifts the short-circuit current slightly away from the direct axis. In the example motor the maximum short-circuit current with resistance taken into account is 28.7 A, of which the negative d-axis component is 28.3 A. With this current applied in the d-axis

$$u_1' = 786 \text{ At}$$

and with the two potentials combined the magnetizing force in the magnet is

$$H_m = -212 \text{ kA/m} = -2.7 \text{ kOe}.$$

This calculation assumed that the magnet remanent flux-density was unaffected by temperature, but it also ignored any local field intensification. The demagnetization curves for NeIGT 24 K show that this magnet would not demagnetize in the example motor even at 182°C.

In line-start motors the maximum demagnetizing current may be two or three times larger than the short-circuit current during the final pole-slips before synchronization. Approximately,

$$I_{d(max)} = \frac{V + E_q}{X_d}$$

where $V$ is the r.m.s. terminal phase voltage.

## 6.3 Steady-state phasor diagram

The steady-state phasor diagram can be constructed in the same way as for the non-salient-pole PM motor in Section 5.3. The open-circuit phase e.m.f. is

$$E_{ph} = jE_q = j\omega \Psi_{Md1}$$

where the subscript Md1 indicates that the flux-linkage is due to the fundamental space-harmonic component of d-axis flux produced by the magnet. In the pure reluctance motor,

$$E_q = 0$$

but there is a d-axis e.m.f. associated with the combined web and link flux:

$$E_{dw} = \omega \Psi_{wq1}.$$

The reactive voltage drops due to armature reaction are constructed as in Fig. 5.5 but with different $X_d$ and $X_q$. The resulting phasor diagram, Fig. 6.8, is described by the two equations

d-axis: $\qquad V_d = E_{dw} - X_q I_q + RI_d$

q-axis: $\qquad V_q = E_q + X_d I_d + RI_q.$

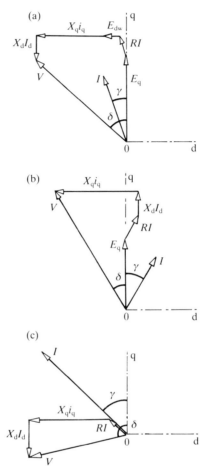

FIG. 6.8. Phasor diagram for PM/reluctance hybrid motor. (a) PM motor with magnetizing armature current in the d-axis. (b) PM motor with demagnetizing armature current in the d-axis. (c) Reluctance motor.

As in Chapter 5, the angles $\delta$ and $\gamma$ are defined as shown, and

$$I_d = -I \sin \gamma \quad V_d = -V \sin \delta$$
$$I_q = I \cos \gamma \quad V_q = V \cos \delta.$$

Positive direct-axis current is magnetizing, and negative is demagnetizing in the PM motor.

The electromagnetic torque is given by

$$T = \frac{3p}{\omega} \left[ (E_q I_q + E_{dw} I_d) + (X_d - X_q) I_d I_q \right].$$

In a two-phase motor the torque is $\frac{2}{3}$ of this value. The torque is thus the sum of 'alignment' torques and reluctance torques. Since $X_d < X_q$, and $I_q$ is positive, $I_d$ must be negative if the reluctance torque is to aid the magnet alignment torque. In a sense these demagnetizing ampere-turns are a price to be paid for the reluctance torque. With negative $I_d$, the torque tending to align the web with the stator m.m.f. is also in the positive direction. Obviously the current phasor needs to be aligned in the second quadrant to maximize the torque per ampere, the power factor, and the efficiency. This is illustrated in Fig. 6.9, which, although schematic, represents a 'snapshot' of the rotating field and stator m.m.f. wave with all torque components positive (anticlockwise).

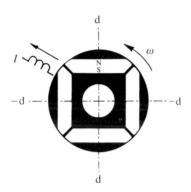

FIG. 6.9. Orientation of stator m.m.f. to produce positive motoring torque.

With no independent source of excitation and $X_d < X_q$, the pure reluctance motor operates with $V$ in the third quadrant and $I$ in the second quadrant, Fig. 6.8(c). Because there are no magnets the stator current has a magnetizing component necessary to sustain the airgap flux, and the power factor is consequently much lower than in the PM motor.

Given the magnitude of the terminal voltage $V$, and its phase relative to the d-axis, we can solve for the d- and q-axis components of phase current:

$$I_d = \frac{X_q(V_q - E_q) + R(V_d - E_{dw})}{R^2 + X_d X_q};$$

$$I_q = \frac{R(V_q - E_q) - X_d(V_d - E_{dw})}{R^2 + X_d X_q}.$$

If $R$ is negligible, as it might be in larger motors,

$$I_d = \frac{V_q - E_q}{X_d}; \quad I_q = -\frac{V_d - E_{dw}}{X_q}.$$

In the pure reluctance motor, $E_q = 0$ and

$$I_d = \frac{X_q V_q + R(V_d - E_{dw})}{R^2 + X_d X_q};$$

$$I_q = \frac{R V_q - X_d (V_d - E_{dw})}{R^2 + X_d X_q}.$$

If $R$ is negligible, as it might be in larger motors,

$$I_d = \frac{V_q}{X_d};$$

$$I_q = -\frac{V_d - E_{dw}}{X_q}.$$

In normal motoring operation $V_d$ and $V_q$ are both negative, and the voltage phasor is in the third quadrant. The current is then in the second quadrant with $I_d$ negative and $I_q$ positive.

To bring out the difference between the PM motor and the pure reluctance motor, we use the motor of Fig. 6.2 as an example. At 3000 r.p.m. the supply frequency is 100 Hz and the reactances and e.m.f. are as shown in Table 6.1. However, the phase resistance is increased by 20 per cent to 0.56 Ω to allow for a temperature rise of about 50°C.

When this motor is operating with a winding current density of 5 A/mm² its phase current is 4.0 A, and this will be taken as the maximum permissible current. We will assume also that the controller is limited to the same current, while its voltage is limited to 38 V r.m.s. This is approximately the maximum a.c. voltage that can be obtained from a 'smartpower' H-bridge rated at 60 V d.c. and operating in the linear mode. The operating conditions to be discussed are calculated from the phasor equations and summarized in Table 6.2.

It is important to note that the operating points in this table are calculated values representing electromagnetic capability, and could be expected to be attained during intermittent operation. The thermally continuous ratings may be only of the order of 50 per cent of those indicated, particularly for the rare-earth motors because of high core losses at high speed. The approach taken admittedly falls short of a thorough comparison but intentionally it compares the basic electromagnetic 'mechanisms' of the different motors in producing torque per ampere and torque per unit volume.

To organize the discussion we will proceed by rows, which are ordered according to a sequence of questions.

*Row 1:*  What is the performance of the PM motor at maximum current, maximum voltage, and 3000 r.p.m.?

With the current oriented 15° ahead of the q-axis the motor develops 287 W and 0.913 N m at a power factor of unity. The torque per ampere is

**Table 6.2.** Operation of PM and synchronous reluctance motors

| * | Motor Parameter | $B_r$ | r.p.m. | $E_q$ | $I$ | $\gamma$ | $I_d$ | $I_q$ | $V_d$ | $V_q$ | $V$ | $\delta$ | $T$ | PF | $P$ | VA/W | $\sigma$ | $T/I$ |
|---|---|---|---|---|---|---|---|---|---|---|---|---|---|---|---|---|---|---|
| Row | Unit | T | | V | A | ° | A | A | V | V | V | degrees | N m | | W | | p.s.i. | N m/A |
| 1 | PMH | 1.1 | 3000 | 35.8 | 4.0 | 15 | −1.04 | 3.86 | −10.12 | 36.73 | 38.10 | 15.40 | 0.913 | 1.000 lg | 287.0 | 1.062 | 0.995 | 0.228 |
| 2 | PMH | 1.1 | 3000 | 35.8 | 4.0 | 0 | 0.0 | 4.0 | −9.88 | 38.04 | 39.30 | 14.56 | 0.912 | 0.968 lg | 286.4 | 1.098 | 0.994 | 0.228 |
| 3 | REL | — | 3000 | — | 4.0 | 45 | −2.83 | 2.83 | −8.58 | −1.76 | 8.75 | 101.6 | 0.066 | 0.551 lg | 20.7 | 3.38 | 0.072 | 0.016 |
| 4 | REL | — | 14 250 | — | 4.0 | 45 | −2.83 | 2.83 | −34.79 | −14.28 | 37.60 | 112.3 | 0.066 | 0.386 lg | 98.1 | 3.07 | 0.072 | 0.016 |
| 5 | REL with $R=0$ and $X_\sigma=0$ | — | 3000 | — | 4.0 | 45 | −2.83 | 2.83 | −5.15 | −1.50 | 5.36 | 106.2 | 0.066 | 0.481 lg | 20.7 | 2.61 | 0.072 | 0.016 |
| 5a | REL with increased slot area and turns | — | 3000 | — | 4.0 | 45 | −2.83 | 2.83 | −35.95 | −11.77 | 37.83 | 108.1 | 0.304 | 0.452 lg | 95.5 | 3.17 | 0.332 | 0.076 |
| 6a | PMH | 0.41 | 3000 | 11.3 | 4.0 | 0 | 0.0 | 4.0 | −9.88 | 13.58 | 16.79 | 36.04 | 0.289 | 0.807 lg | 90.7 | 1.48 | 0.315 | 0.072 |
| 6b | PMH | 0.41 | 3000 | 11.3 | 4.0 | 15 | −1.04 | 3.86 | −10.12 | 12.28 | 15.92 | 39.50 | 0.312 | 0.910 lg | 98.0 | 1.30 | 0.340 | 0.078 |
| 6c | PMH | 0.41 | 3000 | 11.3 | 4.0 | 25 | −1.69 | 3.63 | −9.90 | 11.38 | 15.08 | 41.03 | 0.312 | 0.961 lg | 98.0 | 1.23 | 0.340 | 0.078 |
| 6d | PMH | 0.41 | 3000 | 11.3 | 4.0 | 45 | −2.83 | 2.83 | −8.57 | 9.59 | 12.86 | 41.79 | 0.270 | 0.998 ld | 84.8 | 1.21 | 0.294 | 0.068 |
| 7 | SPM | 1.1 | 3000 | 47.6 | 4.0 | 0 | 0.0 | 4.0 | −4.09 | 49.84 | 50.01 | 4.69 | 1.212 | 0.997 lg | 381.0 | 1.05 | 1.322 | 0.303 |
| 8 | SPM | 1.1 | 2250 | 35.7 | 4.0 | 0 | 0.0 | 4.0 | −3.07 | 37.94 | 38.06 | 4.62 | 1.212 | 0.997 lg | 285.6 | 1.07 | 1.322 | 0.303 |
| 9 | SPM | 1.1 | 3000 | 47.6 | 38.1 | 70 | −35.8 | 13.0 | −33.3 | 18.29 | 38.02 | 61.24 | 3.940 | 0.988 ld | 1238 | 2.34 | 4.300 | 0.103 |
| 10 | SPM | 0.41 | 3000 | 17.9 | 4.0 | 0 | 0.0 | 4.0 | −4.09 | 20.14 | 20.55 | 11.47 | 0.456 | 0.980 lg | 143.2 | 1.148 | 0.497 | 0.114 |
| 11 | SPM | 0.41 | 5850 | 34.9 | 4.0 | 0 | 0.0 | 4.0 | −7.97 | 37.15 | 37.99 | 12.11 | 0.456 | 0.978 lg | 279.2 | 1.089 | 0.497 | 0.114 |
| 12 | SPM | 0.41 | 7500 | 44.8 | 4.0 | 55 | −3.28 | 2.29 | −7.70 | 37.66 | 38.44 | 11.55 | 0.261 | 0.726 ld | 205.3 | 1.498 | 0.285 | 0.065 |
| 13 | PMH | 0.41 | 8400 | 31.8 | 4.0 | 25 | −1.69 | 3.63 | −26.02 | 28.20 | 38.37 | 42.70 | 0.312 | 0.953 lg | 274.5 | 1.118 | 0.340 | 0.078 |
| 14 | PMH | 0.41 | 10 500 | 39.7 | 4.0 | 48 | −2.95 | 2.70 | −25.01 | 29.02 | 38.31 | 40.75 | 0.260 | 0.993 lg | 286.4 | 1.070 | 0.284 | 0.065 |

*See text for discussion
PMH = PM/reluctance hybrid
REL = synchronous reluctance
SPM = surface PM motor
lg = lagging
ld = leading
50°C temperature rise assumed
Volts are a.c. r.m.s.

0.228 N m/A, and the terminal voltage is nearly 38 V. This is taken as the 'base' case for comparisons.

*Row 2:* What is the performance of the PM motor if the current is oriented in the q-axis, as it would normally be for the surface-magnet motor?

With the same current oriented along the q-axis, at the same speed, the torque decreases slightly to 0.912 N m. However, there is a small increase in the required voltage to 39.3 V, indicating that operation at this torque from a 38 V supply requires the field-weakening that results from advancing the current relative to the q-axis.

*Row 3:* What is the performance of the pure reluctance motor at maximum current and 3000 r.p.m.?

The reluctance motor has all the same parameters as the PM motor but $E_q = 0$. The optimum orientation for the current is at $\gamma = 45°$, giving a torque of only 0.066 N m and a specific torque of only 0.016 N m/A with maximum current flowing. However, the voltage required is only 8.75 V, implying that the reluctance motor could sustain this torque up to a higher speed. Note the very low power factor, 0.55 lagging, and the high volt-ampere requirement.

*Row 4:* Up to what speed could the reluctance motor sustain its maximum torque per ampere?

By solving the phasor equations with $I = 4.0$ A, $V = 37.6$ V, and $\gamma = 45°$, assuming that all the reactances are proportional to frequency, the frequency is found to be approximately 475 Hz and the speed 14 250 r.p.m. The power factor becomes even lower. Note that the 'volts per Hz' remained roughly constant in this calculation.

Evidently the reluctance torque is only about 7 per cent of the total torque of the PM motor, with the same voltage and current, but it must be borne in mind that these are small motors in which the resistance and leakage reactance are comparatively large. In a large motor both of these impedances would be smaller in relation to the synchronous reactances. An estimate of the expected relative improvement in larger motors can be made by assigning the resistance and leakage reactance to zero. While this is artificial, it permits all the other parameters to be kept the same, and provides a kind of 'per-unit' comparison.

*Row 5:* Reluctance motor with $R = X_\sigma = 0$

With 4.0 A oriented at 45° ahead of the q-axis, the torque remains the same because it depends on the difference between $X_d$ and $X_q$; the leakage reactance is common to both. However, the terminal voltage is now reduced to 5.36 V. The power factor actually decreases to 0.481 lagging, because of the removal of the in-phase voltage drop across the phase resistance. The efficiency of this hypothetical motor is 100 per cent, and even with no losses the volt-amperes per watt are still more than twice the value needed by the PM motor. At constant volts/Hz the torque could be maintained constant up to a speed of $38 \div 5.36 \times 3000 = 21\ 270$ r.p.m. without exceeding the current limit.

*Row 5a:* Reluctance motor with increased slot area and turns

The low voltage required by the reluctance motor is so far below the available converter voltage that the system comparison is not quite fair to the reluctance motor. Although the copper loss is the same as in the PM motors, the core loss must be much lower. To correct this, row 5a shows results for the reluctance motor with double the slot area and 2.15 times the turns. With the same slot-fill factor and a slightly smaller wire size, the current density is increased by 7.5 per cent and with 4.0 A of phase current the copper losses are increased by 2.3 times to 41.4 W, which is of the same order as the total loss in the PM motors that use high-energy magnets. The core losses are assumed to remain very low. The reactances increase to 5.45 Ω and 11.42 Ω and the resistance to 1.29 Ω at 100 Hz. With 4.0 A oriented at 45°, the motor now requires a terminal voltage of 37.8 V. Row 5a shows the improvement in performance, with the torque increased from 0.066 N m to 0.304 N m, an improvement of 4.6 times. The torque per ampere, 0.076 N m/A, is comparable with that of the ceramic-magnet hybrid motor (rows 6 a–d). The power factor remains low and the volt-ampere requirement remains high. Also, the maximum speed will be reduced by a factor of the same order as the increase in the torque per ampere. Thus the wide speed range has been completely sacrificed to get a 'competitive' value of the torque per ampere.

*Row 6:* PM motor with ceramic magnets at maximum current, 3000 r.p.m. This calculation assumes a remanent flux-density of 0.41 T. The corresponding open-circuit e.m.f. is calculated by the method of Section 6.2, assuming the links to be saturated. Thus $E_q = 11.3$ V. Operation is shown with four different advance angles for the current phasor, in rows 6a–d. The highest torque per ampere is obtained with $\gamma = 25°$ at a power factor of 0.961 lagging. The specific torque is 0.078 N m/A. Even with the 'penalty' of a large per-unit leakage reactance and significant resistance, the ceramic PM motor still outperforms the pure reluctance motor with nearly five times the specific torque. It needs only 15.1 V terminal voltage, suggesting that at constant volts/Hz the speed could be raised to about 7500 r.p.m. at constant torque, without exceeding the current limit.

It is evident from these calculations that the weaker the d-axis magnetic field, the wider the speed range. The PM motor has the capability to weaken the field while maintaining a high power factor, operating with a current lead angle relative to the q-axis. The reluctance motor is the ultimate case of a field-weakened motor.

*How do these motors compare with the surface-magnet motor?*
If we take the same stator and the same magnet volume distributed in four poles with the same pole-arc (136 electrical degrees), the magnet length in the radial direction is 5.15 mm and with the same remanent flux-density (1.1 T) and the same resistance and leakage reactance, $E_q = 47.6$ V and $X_s = 1.022$ Ohm. The airgap flux-density is rather high at 0.97 T, which would produce unacceptable saturation levels in the stator teeth and increase the core losses, but these factors are left out of consideration while we examine the other basic effects of putting the magnets on the rotor surface.

*Row 7:* Surface-magnet motor at maximum current and 3000 r.p.m.

Assuming that the current is oriented along the q-axis ($\gamma = 0$), the torque is 1.21 N m, giving 381 W of shaft power at a specific torque of 0.303 N m/A and a power factor of 0.997 lagging. The specific torque exceeds that of the hybrid PM/reluctance motor by 33 per cent. However, the terminal voltage needed to achieve it is 50 V, which far exceeds the assumed controller limit.

*Row 8:* What is the highest speed at which the surface-magnet motor can maintain maximum torque within the current and voltage limits?

At constant volts/Hz, the maximum speed would be $38 \div 50 \times 3000 = 2280$ r.p.m. with a frequency of 76 Hz. A more accurate calculation gives 75 Hz. At this frequency $E_q$ is reduced to 35.7 V. With $\gamma = 0$ and $I = 4.0$ A, the torque is 1.21 N m as before, and the power factor is the same.

*Row 9:* Could the surface-magnet motor be operated at 3000 r.p.m. at all, with maximum controller voltage of 38 V?

To operate the surface-magnet motor at this voltage and frequency requires a massive demagnetizing current of 35.8 A in the d-axis, giving a total current of 38.1 A, far in excess of the assumed limit and well above even the short-time rating of the windings. Therefore it is not possible to operate the surface-magnet motor at 3000 r.p.m. within the prescribed voltage and current limits. This emphasizes the point made in Chapter 5, that the surface-magnet motor is not naturally adapted to field-weakening operation.

It might be argued that the surface-magnet motor could be re-wound to operate at 38 V and 100 Hz (3000 r.p.m.); but to produce the same torque the current would increase in inverse proportion to the decrease in the number of turns. The controller would have the same voltage, but a higher current rating, and therefore bigger devices would be needed. If high-speed operation is only for a short period, however, there may be cases where the original devices could provide the increased current within their short-duration thermal ratings.

*Row 10:* Surface-magnet motor with ceramic magnets at maximum current and 3000 r.p.m.

The remanent flux-density is again assumed to be 0.41 T for the ceramic magnets, but otherwise the design is the same as before. With $E_q = 17.9$ V and $I = 4.0$ A oriented along the q-axis, the torque is 0.456 N m at 0.114 N m/A, giving a shaft power of 143 W at a power factor of 0.980 lagging. The specific torque exceeds that of the ceramic hybrid motor by 46 per cent. The terminal voltage needed is 20.6 V, suggesting that at constant volts/Hz the torque could be maintained constant up to a speed of $38 \div 20.6 \times 3000 = 5500$ r.p.m. without exceeding the current limit. This compares with the expected 7500 r.p.m. of the hybrid motor.

*Row 11:* What is the maximum speed at which the ceramic surface-magnet motor can develop maximum torque?

With all the current in the q-axis the required terminal voltage reaches 38 V at 5850 r.p.m. with a frequency of 195 Hz. The torque is still 0.456 N m as it was at 3000 r.p.m. and 20.6 V, and the power output is almost doubled at 279 W.

*Row 12:* What happens to the surface-magnet motor at higher speeds?
The torque falls off rapidly. At 7500 r.p.m. (250 Hz), with the controller at its maximum voltage, maximum current can still be forced into the motor but only with $\gamma = 55°$. The large d-axis current is needed to force the induced voltage down to the maximum level of the controller, and the power factor angle is 43.5° leading. The torque has fallen off to 0.261 N m and the specific torque to only 0.065 N m/A, while the power is down to 205 W. The motor is not even maintaining constant power at these higher speeds.

*Row 13:* What is the maximum speed at which the hybrid PM/reluctance ceramic magnet motor can maintain maximum torque?
At 8400 r.p.m. (280 Hz) the hybrid motor can operate with $\gamma = 25°$ to produce 0.312 N m, the same as at 3000 r.p.m. The maximum speed at maximum torque is close to this value, and is thus about $8400 \div 5700 = 1.47$ times higher than the corresponding speed for the surface-magnet motor. It is no coincidence that this is the same ratio as the torque per ampere of the two motors, and indeed their behaviour in this regard is exactly the same as a d.c. motor with field weakening. The ratio is smaller when high-energy magnets are used, because the reluctance torque and the per-unit reactances are then smaller, and the hybrid motor assumes more of the character of the surface-magnet motor.

*Row 14:* Constant-power operation of the hybrid motor
At 10 500 r.p.m. (350 Hz) the hybrid motor requires a significant degree of field weakening to constrain its voltage. Accordingly it must operate with a greater value of demagnetizing current in the d-axis and a smaller q-axis current. The shift of $\gamma$ to larger values reduces the magnet alignment torque but simultaneously increases the reluctance torque until $\gamma$ reaches 45°. Operation in Row 14 shows that while the torque has fallen to 0.261 N m, the power has actually increased slightly to 286.4 W from the 274.5 W at 8400 r.p.m.

These calculations ignore the effects of core losses, which become greatly increased at higher speeds. However, the constant-power characteristic of the hybrid motor has been well established experimentally (Jahns 1988; Jahns et al 1987).

*Summary* The hybrid PM/reluctance motor has lower torque per ampere than the surface-magnet motor, given the same volume of the same magnet material and the same stator. With a remanent flux density of 1.1 T its specific torque is 0.228 N m/A compared with 0.303 for the surface-magnet motor. However, the higher figure can be achieved only with high flux-densities in the stator teeth, and when core losses are taken into account, the useable torque may be considerably less, and so may the efficiency. With ceramic magnets the specific torque of the surface-magnet motor is 47 per cent more than that of the hybrid, suggesting that the surface-magnet configuration makes more effective use of low-energy magnets than does the hybrid motor.

For the same controller volt-ampere rating, the hybrid motor can maintain its maximum torque over a wider speed range than the surface-magnet motor. Comparing the high-energy magnet motors, the hybrid can work at maximum torque at speeds up to 3000 r.p.m., whereas for the same controller volt-amperes the surface-magnet motor is limited to 2250 r.p.m. Comparing the ceramic magnet motors, the hybrid can work at maximum torque up to 8400 r.p.m. compared with only 5850 r.p.m. for the surface-magnet motor. The hybrid motor can also maintain constant power up to about 10 500 r.p.m.

The specific torque and the torque per ampere of the pure synchronous reluctance motor are comparable with those of the ceramic-magnet hybrid motor, but only if the stator is redesigned to accommodate about twice as much copper and twice the ampere-turns. Copper losses still predominate, except at high speeds. The power factor is extremely low, and this appears to be an inherent characteristic that does not improve with physical size. Apart from its simplicity and ruggedness, its main operational advantage is a wide speed range, albeit with a much lower torque capability than the PM motors.

What other advantages can be adduced for the pure reluctance motor? The absence of magnets indicates a capability to survive very high temperatures, within the limits set by the bearings and the winding insulation. Because the reluctance torque is independent of the direction of the current, it is possible to operate with unipolar currents, and this permits the windings to be connected in series with the converter phaselegs as in the case of the switched reluctance motor (Chapter 7). Such converters are less susceptible to short-circuit faults across the d.c. supply, but on the other hand they cannot use devices that are packaged as half-bridges.

### 6.3.1 Converter volt-ampere requirements

The volt-amperes per watt of shaft power quoted in Table 6.2 are sinewave values and represent the total apparent power at the motor terminals. Elsewhere we have defined the volt-amperes per watt in terms of the volt-ampere product required in the ratings of the semiconductor devices in the converter. The example calculations are all based on a two-phase motor and it was assumed that each phase was supplied by a full bridge circuit, requiring a total of eight transistors. The nominal converter volt-amperes, based on r.m.s. current in each device times peak voltage times the number of devices, is therefore

$$V\sqrt{2} \times 8 \times \frac{I}{\sqrt{2}} = 8VI.$$

The same overall figure would result if a single full bridge was used to supply both windings connected in a centre-tap arrangement.

If a three-phase motor were used, the terminal volt-ampere requirements would be the same, but with a three-phase bridge converter the number of devices per phase is only two instead of four. Consequently the total device volt-ampere requirement is only $6VI$. With a nominal a.c. apparent power requirement of about 1.1 VA/W, raised to perhaps 1.2 to allow for core losses and friction, the rough average requirement of both the hybrid and the surface-magnet motors can be reckoned as about 7.2 kVA/kW based on r.m.s. current, and about 10 kVA/kW based on peak current, assuming a three-phase motor.

## 6.4 Circle diagram and torque/speed characteristic

In the complex phasor diagram the maximum continuous phase current defines a circular locus (see Figs 6.10 and 6.11):

$$I_d^2 + I_q^2 = I_c^2.$$

The maximum converter voltage limits the current to a different locus defined by

$$V_d^2 + V_q^2 = V_c^2.$$

If resistance is neglected this can be written

$$X_q^2 I_q^2 + (E_q + X_d I_d)^2 = V_c^2$$

or, equivalently,

$$I_d = \frac{V_c \cos \delta - E_q}{X_d};$$

$$I_q = -\frac{V_c \sin \delta}{X_q}$$

which is an ellipse. If resistance is kept in the reckoning the locus is only approximately elliptical. As the frequency and speed increase (Figs 6.10 and

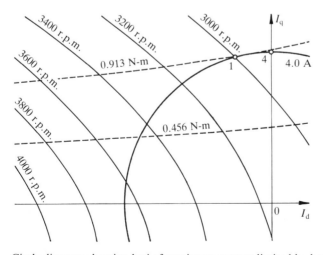

FIG. 6.10. Circle diagram showing loci of maximum current limited by both current and voltage, for hybrid motor with high-energy magnets (1.1 T; V = 38 V).

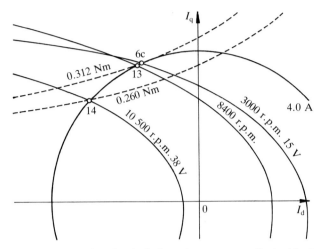

FIG. 6.11. Circle diagram showing loci of maximum current limited by both current and voltage, for hybrid motor with ceramic magnets (0.41 T; V = 15 V and 38 V).

6.11), the voltage-limited current loci decrease in size, just as they do for the surface-magnet sinewave motor, for which the loci are circles. Up to a certain speed it is still possible to get maximum current into the motor, but $I_q$ decreases while the negative (demagnetizing) component $I_d$ increases. The torque decreases and the magnet operating point moves further down its demagnetization characteristic. In Fig. 6.10 the maximum speed at which maximum current can just be forced into the motor is slightly under 3700 r.p.m. with the controller at maximum voltage (38 V), but the torque is zero or negative. At 3000 r.p.m. the current can be oriented anywhere between the q-axis and about 25° ahead. Point 1 corresponds to Row 1 of Table 6.2.

A constant-torque contour can be drawn by rewriting the torque equation:

$$T = \frac{3p}{\omega} \Delta X I_q I'_d$$

(with 3 replaced by 2 in a two-phase motor). Here

$$\Delta X = X_d - X_q \text{ and } I'_d = I_d + \frac{E_q}{\Delta X}.$$

This is a rectangular hyperbola asymptotic to the negative d-axis and to a q-axis offset to the right. Note that all these relationships are independent of frequency and speed. With high-energy magnets the offset $E_q/\Delta X$ is so large that the constant-torque contours are almost horizontal straight lines, as they are for the surface-magnet motor. This again shows the similarity between the two machines when high-energy magnets are used. When the hybrid motor is 'underexcited', as it may well be with ceramic magnets, the constant-torque contours have more curvature, as in Fig. 6.11. For the pure synchronous reluctance motor the constant-torque contours are also rectangular hyperbolas but with no offset.

The torque contour for 0.312 N m in Fig. 6.11 is tangent to the maximum-current circle at point 1, which corresponds to Row 1 in Table 6.2. This torque is attainable at 3000 r.p.m. with a controller voltage of only 15 V. As the speed increases the size of the voltage-limited current locus can be maintained by increasing the voltage (by p.w.m. control) up to the maximum of 38 V, which is reached at 8400 r.p.m. (point 13). This is the highest speed at which the torque of 0.312 N m can be attained, giving an electromagnetic power of 274.5 W at the airgap. If the speed is raised to 10 500 r.p.m. the torque must decrease as the operating point is constrained by the maximum current limit. At point 14, $\gamma = 48°$ and the torque has fallen to 0.261 N m, but the power is slightly higher at 286.4 W. Further increases of speed will cause the power to fall quite rapidly. If the 'base' speed is taken as 8400 r.p.m., that is, the maximum speed at which maximum torque can be developed with maximum available converter voltage and current, then the speed range at constant power is roughly $10\,500 \div 8400 = 1.25$.

The torque and power calculations in this discussion are, of course, optimistic because of the neglect of iron losses and friction and windage.

The torque contours for the three motors show that for the surface-magnet motor, maximum torque per ampere is obtained with $\gamma = 0$; for the pure reluctance motor, $\gamma = 45°$; and for the hybrid motor, $\gamma$ is at some intermediate angle that is closer to 0 the higher the magnet energy product. The control of the hybrid motor is thus more complex than that of either of the other two, particularly as the speed is increased into the constant-power range. The lead angle $\gamma$ must be increased as the speed increases, otherwise the current limit will be exceeded (Jahns 1987).

## 6.5 Cage-type motors

The motor in Fig. 6.3 is a 'line-start' motor; that is, it can start if connected to an a.c. voltage supply because the cast aluminium cage winding produces an asynchronous torque, just as in the induction motor. PM versions of this machine are also possible and are manufactured in relatively small numbers, particularly for applications where it is advantageous to operate several motors in synchronism from a single voltage-source inverter (Jordan 1983). In this case there is no shaft position feedback and the cage acts as an amortisseur to stabilize the operation and prevent oscillations that might otherwise occur in certain speed ranges.

More recently the line-start PM motor has been developed for constant-speed applications where very high efficiency and power factor are needed to minimize electricity costs. In Cornell (1983) an example of a 25 hp, 1800 r.p.m. motor showed a reduction in losses of 34 per cent compared with a high-efficiency induction motor. The space needed to accommodate the starting cage compromises the magnet geometry to some degree. The cage requires special attention to ensure adequate pull-in (synchronizing) torque as well as low-speed or breakaway torque, and a double-cage arrangement may be necessary. It is inherent in the starting of these machines that several pole-slips occur before synchronization, causing severe torque pulsations with many more torque reversals than is the case with induction-motor starts, where the torque pulsations are caused by transient d.c. offsets that die away very quickly. However, such machines are of interest more for applications requiring continuous non-stop operation with infrequent stops and starts. The pull-in (synchronizing) torque is an inverse function of the load inertia (Miller 1984) and it is generally not as high as for a comparable induction motor. For low or moderate inertia loads, however, the synchronizing capability is quite adequate.

## Problems for Chapter 6

1. Show that with rated phase current $I$, the angle $\gamma$ which maximizes the torque of the hybrid PM/reluctance motor satisfies the relationship

$$\frac{\sin \gamma}{\cos 2\gamma} = \frac{\Delta X I}{E_q}$$

where $\Delta X = X_q - X_d$. Hence show that for the motor of Table 6.2 at 3000 r.p.m. with rated current of 4.0 A, the torque is maximized with $\gamma = 7.97°$. What is this maximum torque? Neglect losses.

2. If the high-energy magnets in the motor of Problem 1 are replaced by ceramic magnets, the reactances remain the same but the open-circuit phase e.m.f. is decreased to 11.3 V. Determine the optimum value of $\gamma$ and the maximum torque, assuming that the rated current remains 4.0 A.

3. The absolute maximum speed of the PM/reluctance hybrid motor can be estimated by assuming that the current is at its rated value and is entirely in the negative d-axis. If the maximum controller voltage is $V_c$ then, neglecting losses,

$$V_c = E_q + X_c I$$

where $I_d$ is negative. If $V_c = 38$ V and $E_q = 35.8$ V at 3000 r.p.m., and if $X_d = 1.18\ \Omega$ at 3000 r.p.m., estimate the absolute maximum speed with $I = 4.0$ A, using the fact that $E_q$ and $X_d$ are proportional to frequency or speed (see Fig. 6.10).

4. Re-work Row 1 of Table 6.3 with a phase current of 8.0 A oriented at $\gamma = 15°$.

5. Calculate the synchronous reactances $X_d$ and $X_q$ of the reluctance motor of Table 6.2 (with $E_q = 0$) if the pole–arc/pole–pitch ratio is reduced to $\tfrac{2}{3}$ while keeping all other relevant dimensions the same. Hence calculate the maximum electromagnetic torque with a phase current of 4.0 A.

# 7 Switched reluctance drives

## 7.1 The switched reluctance motor

The concept of the switched reluctance motor was established by 1838 (see Byrne *et al.* 1986), but the motor could not realize its full potential until the modern era of power electronics and computer-aided electromagnetic design. Since the mid-1960s these developments have given the SR motor a fresh start and have raised its performance to levels competitive with d.c. and a.c. drives and brushless d.c. drives.

It is difficult to be certain about the origin of the term 'switched reluctance', but one of the earliest occurrences is in Nasar (1969) in relation to a rudimentary disk motor employing switched d.c. Professor Lawrenson (1980) was perhaps the first to adopt the term in relation to the radial-airgap motor which is the focus of attention today, but the terms 'brushless reluctance motor', 'variable reluctance motor', and 'commutated reluctance motor' are among several equally acceptable alternatives that were in use long before this time. It could perhaps be most accurately described as a 'statically commutated doubly-salient vernier reluctance motor'.

Apart from the well-known work by Lawrenson and his colleagues at the University of Leeds, UK and subsequently at Switched Reluctance Drives Ltd, there have been many other substantial contributions to the technology since the mid-1950s. Among the most notable are the works of French and Williams (1967); GE (Bedford 1972, and several other authors subsequently); Ford (Unnewehr and Koch 1974); Professor J. V. Byrne of University College, Dublin (1972, 1976, 1982, 1986); the Jarret Company in France; Inland-Kollmorgen (Ireland); Professor M. R. Harris of the University of Newcastle upon Tyne; and Professor J. H. Lang of the Massachusetts Institute of Technology.

By the time of Unnewehr and Koch's paper in 1974, most of the basic design principles of both the motor and the control were well understood, but modern technology has facilitated much refinement since that time.

With the exception of large-diameter 'direct drive' robot motors, the only commercially produced SR drives at the time of writing are the Oulton drive (Tasc Drives Ltd) in the UK and a computer plotter servo (Hewlett-Packard/Warner Electric) in the US. However, this situation is likely to change rapidly as the technology matures and becomes more widely understood.

The switched reluctance motor is a doubly-salient, singly-excited motor. This means that it has salient poles on both the rotor and the stator, but only one member (usually the stator) carries windings. The rotor has no windings, magnets, or cage winding, but is built up from a stack of salient-pole laminations, see Fig. 7.1(a–f).

There are two essentials that distinguish the SR motor from the variable-reluctance stepper (Kenjo, 1985). One is that the conduction angle for phase currents is controlled and synchronized with the rotor position, usually by means of a shaft position sensor. In this respect the SR motor is exactly like the PM brushless d.c. motor, but unlike the stepper motor, which is usually fed with a squarewave of phase current without rotor position feedback. The second distinction between SR and stepper motors is that the SR motor is designed for efficient power conversion at high speeds comparable with those of the PM brushless d.c. motor; the stepper, on the other hand, is usually designed as a torque motor with a limited speed range. Although this may seem a fine distinction, it leads to fundamental differences in the geometry, power electronics, control, and design technique.

The SR motor is more than a high-speed stepper motor. It combines many of the desirable qualities of both induction-motor drives and d.c. commutator motor drives, as well as PM brushless d.c. systems. Its performance and inherently low manufacturing cost make it a vigorous challenger to these drives. Its particular advantages may be summarized as follows:

(1) The rotor is simple and requires relatively few manufacturing steps; it also tends to have a low inertia.

(2) The stator is simple to wind; the end-turns are short and robust and have no phase–phase crossovers.

(3) In most applications the bulk of the losses appear on the stator, which is relatively easy to cool.

(4) Because there are no magnets the maximum permissible rotor temperature may be higher than in PM motors.

(5) The torque is independent of the polarity of phase current; for certain applications this permits a reduction in the number of power semiconductor switches needed in the controller.

(6) Under fault conditions the open-circuit voltage and short-circuit current are zero or very small.

(7) Most converter circuits used with SR motors are immune from shoot-through faults, unlike the inverters used with a.c. and brushless d.c. drives.

(8) Starting torque can be very high, without the problem of excessive inrush currents, as for example the large starting current of induction motors at high slip.

(9) Extremely high speeds are possible.

(10) The torque/speed characteristic can be 'tailored' to the application requirements more easily than in the case of induction motors or PM motors.

These are clear advantages that require little or no qualification. Other advantages must be seen in the light of the trade-offs that go with them. Because there is no fixed magnet flux the maximum speed at constant power is

# THE SWITCHED RELUCTANCE MOTOR

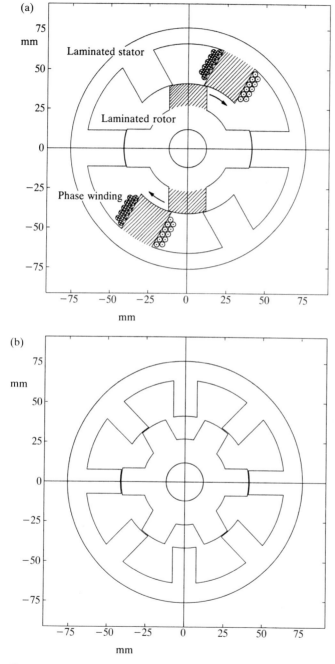

FIG. 7.1. Cross-sections of switched reluctance motors generated by CAD program. One phase comprises windings on opposite poles. (a) Three-phase 6:4. (b) Four-phase 8:6.

FIG. 7.2. (a) A 700 W, 10 000 r.p.m. SR motor (right) from Switched Reluctance Drives Ltd., and a conventional commutator motor (left) of the same performance but 2.5 times the volume.

FIG. 7.2. (b) The components for an 800 W, 240 V integrated controller for a four-quadrant SR drive, power module, capacitor, rectifier, and interface ICs. Courtesy Switched Reluctance Drives Ltd., Leeds, UK.

not as restricted by controller voltage as it is in PM motors. However, the absence of 'free' PM excitation imposes the burden of excitation on the stator windings and the controller, and increases the per-unit copper losses. Particularly in small motors this is a disadvantage that limits the efficiency and the torque per ampere.

The SR motor also has some clear disadvantages. The most important is the pulsed, or at least non-uniform, nature of the torque production which leads to torque ripple and may contribute to acoustic noise. Over a narrow speed range it is possible to reduce the torque ripple to less than 10 per cent r.m.s., which is comparable with the levels attainable in induction motors and other brushless d.c. drives, but it is practically impossible to maintain this level of smoothness over a wide speed range. Fortunately it is easier to achieve smooth torque at low speeds, where many loads are most sensitive to torque ripple effects. The acoustic noise can be severe in large machines where ultrasonic chopping frequencies are not practical. But even in small ones, when all steps have been taken to minimize chopper noise, there remains a characteristic sound similar to 'tickover' noise in internal combustion engines at light load; under heavy load this tends to become a 'growl' that may be difficult to eliminate. The noise level is sensitive to the size, being much less severe in small machines. It also depends on the mechanical construction and the precision of the firing angles. The torque ripple is also sensitive to these factors. Although the construction is simple, electrical and mechanical precision are essential to keep it quiet and this tends to increase the cost.

A further aspect of the torque ripple is that the ripple current in the d.c. supply tends to be quite large, making for a large filter capacitance requirement. This in turn may cause significant a.c. line harmonics in systems operating from rectified a.c.

The SR motor makes use of the 'vernier' principle common in stepper motors, in which an internal torque multiplication is achieved with a rotor speed slower than that of a rotating-field machine with the same number of phases and rotor poles. Without this multiplication the torque per unit volume would be much less than that of induction motors and PM motors, and the price paid is a substantial increase in commutation frequency, which may lead to higher core losses and converter switching losses. The effect is compensated by a smaller volume of iron than that of a comparable a.c. motor, and also by the fact that in some sections of the core the flux excursions are unipolar, which helps to limit the hysteresis losses. The internal torque multiplication or vernier effect compensates for the relatively poor utilization of converter volt-amperes and restores the effective power factor and torque per ampere to a 'competitive' level at the expense of switching frequency and magnetic losses. This mechanism is not applicable to the synchronous reluctance motor, and this is the main reason why its weak performance relative to its PM stablemates does not carry over to the switched reluctance motor.

For optimum performance the airgap needs to be about the same as that of an induction motor of comparable diameter, or perhaps slightly larger; PM brushless motors, however, can operate with larger airgaps and therefore slightly larger manufacturing tolerances.

The pole shape of an SR motor cannot be made 'square' as is the normal tendency in a.c. and d.c. machines. This is true of virtually all known designs of

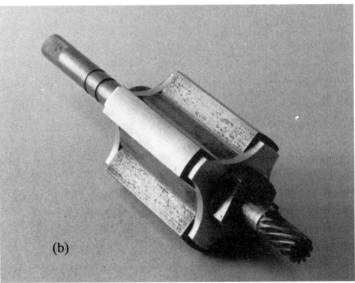

FIG. 7.3. Components of a small SR motor designed for general-purpose variable-speed gearmotor applications in the 100–400 W range. (a) Stator. Note the essential simplicity of construction and windings. This motor, designed for a highly noise-sensitive application, has compression fingers bearing on the stator poles. The short, robust endwindings are also evident, with no phase crossovers. Cooling air has excellent access all along the stator slots. The slot fill in this motor is less than 30 per cent. (b) Rotor. Comprising only a stack of laminations, the SR rotor is the simplest of all electric motor rotors. The ground surface of the rotor poles helps to achieve a uniform airgap, as in induction motors. Windage loss is not as great as the salient-pole construction might suggest; but the air turbulence helps to cool both the rotor and the stator. Courtesy Scottish Power Electronics and Electric Drives, Glasgow, UK.

SR motor, yet very little analysis has appeared in the literature, even though it undoubtedly has a profound effect on its characteristics relative to those of more conventional machines. Long narrow poles tend to produce the best designs by reducing the effects of end-winding inductance and resistance; but this also has the effect of reducing the flux and inductance, and the SR motor typically requires more turns of thinner wire than an a.c. motor wound for the same voltage. In small drives with a wide speed range, this tends to require a lower minimum duty-cycle in the chopping of the supply voltage, and if the chopping frequency is high, special high-frequency pulse techniques and very fast power switches and diodes may be necessary. However, the same may be true of the surface-magnet brushless d.c. PM motor, which also has a low inductance.

Much effort has been expended in attempts to compare the power output and efficiency of SR motors with those of competing drive technologies. For any such comparison to be meaningful it is necessary to restrict it to a narrow set of specifications. It is impossible to make completely general statements about relative performance; far too many variable parameters are involved. There are surprisingly few detailed comparisons in the literature. In larger sizes it is likely to be found that when all aspects of the performance are put on an equal basis, the SR motor is no smaller than an induction motor designed to the same specification. In small sizes the power density, or equivalently the efficiency, of both these motors falls off and neither of them can attain the performance of the brushless PM motor.

The SR motor cannot start or run from an a.c. voltage source, and it is not normally possible to operate more than one motor from one inverter. It is normally necessary to use a shaft position sensor for commutation and speed feedback. Serious attempts have been made to operate without the sensor but inevitably there is a price to be paid either in performance or in control complexity.

The cabling for SR motors is typically more complex than for induction motor drives: a minimum of four wires, and more usually six, are required for a three-phase motor, in addition to the sensor cabling.

This somewhat lengthy review of the disadvantages of SR motors is included in the interests of making a balanced appraisal of it as it stands today. The weighting attached to each advantage and each disadvantage is different for every application, and the weighted sum can only be evaluated relative to a detailed specification. In view of the distinct characteristics of the SR drive, its likely application is where a brushless drive is required with a wide speed range, with a cost saving over the conventional PM brushless d.c. motor drive. The control is simpler than the field-oriented induction-motor drive but in larger machines this does not necessarily mean that it will be less expensive, because control costs for a given level of functionality are tending to decrease. The noise and torque ripple are likely to remain worse than brushless d.c. PM motors, but in small drives this may not be a particular problem. By comparison with small commutator motors (a.c. or d.c.), the SR motor can fairly be claimed to be significantly quieter.

## 7.2 Poles, phases, and windings

The 'classical' forms of switched reluctance motor are those in Fig. 7.1, with stator:rotor pole numbers of 6:4 and 8:6. Others are possible, including 4:2, 6:2, 10:4, 12:8, and variants with more than one tooth per pole such as 12:10, (Finch 1984). Only the two shown in Fig. 7.1 are considered here.

Many of the basic rules constraining the choice of pole numbers, pole arcs, and phase number were expounded by Lawrenson (1980). The relationship between speed and fundamental switching frequency follows from the fact that if the poles are wound oppositely in pairs to form the phases, then each phase produces a pulse of torque on each passing rotor pole; the fundamental switching frequency in one phase is therefore

$$f_1 = nN_r = \frac{\text{r.p.m.}}{60} N_r \text{ Hz}$$

where $n$ is the speed in rev/s and $N_r$ is the number of rotor poles. If there are $q$ phases there are $qN_r$ steps per revolution and the 'step angle' or 'stroke' is

$$\varepsilon = \frac{2\pi}{qN_r} \text{ rad.}$$

The number of stator poles usually exceeds the number of rotor poles.

The pole arcs are determined by the essential torque-production mechanism, which is the tendency of the poles to align. If fringing is neglected there must be overlap between a pair of rotor poles and the poles of the excited stator phase; in this case torque can be produced through an angle $\beta$, which is the smaller of the stator and rotor pole arcs. To produce unidirectional torque through 360° it is obvious that $\beta$ must not be smaller than the step angle, otherwise there will be 'gaps' where no torque is produced: thus

$$\beta > \varepsilon.$$

In order to get the largest possible variation of phase inductance with rotor position, the interpolar arc of the rotor must exceed the stator pole arc. This leads to the condition

$$\frac{2\pi}{N_r} - \beta_r > \beta_s$$

which ensures that when the rotor is in the 'unaligned' position relative to the stator poles of one phase, there will be no overlap and therefore a very low inductance. The unaligned position is defined as the conjunction of any rotor interpolar axis with the axis of the stator poles of the phase in question. In Fig. 7.1(b) the phase on the vertical axis is 'unaligned' while the phase on the horizontal axis is 'aligned'.

A further constraint on the pole arcs is that usually the stator pole arc is made slightly smaller than the rotor pole arc. This permits slight increases in the slot area, the copper winding cross-section, and the aligned/unaligned inductance ratio.

The constraints on pole arcs can be expressed graphically as in Fig. 7.4, in which the 'feasible triangles' (Lawrenson, 1980) define the range of combinations normally permissible. As might be expected, the variation in performance of machines defined by different points in these triangles is

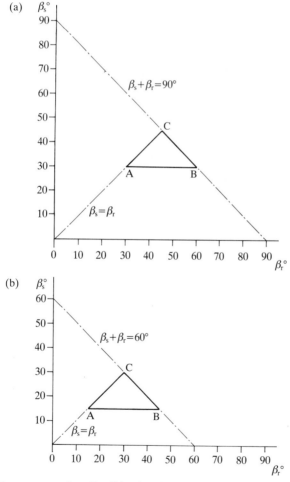

FIG. 7.4. Pole-arc constraints (feasible triangles). (a) Three-phase 6:4. (b) Four-phase 8:6. (Lawrenson 1980).

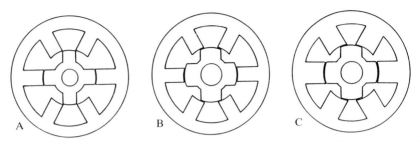

FIG. 7.5. Cross-sections obtained from the vertices of the feasible triangles of Fig. 7.4(a). The relative proportions determine the inductance ratio and the copper and core losses.

considerable. Figure 7.5 shows, for a three-phase motor, the cross-sections corresponding to the vertices A, B, and C on Fig 7.4(a). Design C is likely to have too high an unaligned inductance and too little winding area. Design B has more copper area but still the unaligned inductance will be high because of fringing. Design A has a large winding area and a high inductance ratio, leading to a high efficiency and power density, but its torque ripple is higher than in the others.

The 'optimum' tooth-width/tooth-pitch ratio used in stepper motor design is not applicable to the SR motor. It is of course possible to determine a combination of pole arcs that gives the highest inductance ratio and therefore the highest 'static torque per ampere'. But too many other factors have to be considered to make this the universal choice. Among them are the torque ripple, the starting torque, and the effects of saturation. Curvature effects are also more pronounced than in steppers because of the small number of poles. As in steppers, pole taper is likely to be of benefit in reducing core losses, the m.m.f. drop in the rotor and stator steel, and the adverse effects of saturation. Stator pole taper also reduces the unaligned inductance, but it slightly decreases the winding area.

Several other detailed modifications to the simple geometry of Fig. 7.5 are permissible and advantageous, such as the use of a hexagonal stator blank which increases the winding area and can produce a mechanically stiffer core, at the same time reducing the scrap from the punching process. Pole overhangs can be used to control the local saturation during the initial overlap period. Welding the outside of the stator stack is permissible as in a.c. motors, but this cannot be used on the rotor, which requires mechanical means for compressing the laminations together. The rotor may be skewed slightly to reduce noise.

## 7.3 Static torque production

Consider the primitive reluctance motor in Fig. 7.6(a). When current is passed through the phase winding the rotor tends to align with the stator poles; that

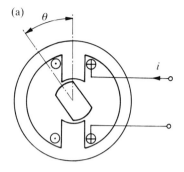

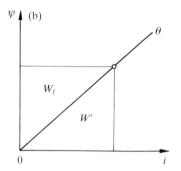

FIG. 7.6. Elementary reluctance motor showing principle of torque production: (a) primitive motor; (b) field energy and coenergy.

is, it produces a torque that tends to move the rotor to a minimum-reluctance position.

In a device of this type, the most general expression for the instantaneous torque is

$$T = \left[\frac{\partial W'}{\partial \theta}\right]_{i=\text{constant}}$$

where $W'$ is the coenergy defined as in Fig. 7.6(b):

$$W' = \int_0^i \psi \, di.$$

An equivalent expression is

$$T = -\left[\frac{\partial W_f}{\partial \theta}\right]_{\psi=\text{constant}}$$

where $W_f$ is the stored field energy defined as in Fig. 7.6(b):

$$W_f = -\int_0^\psi i \, d\psi.$$

When evaluating the partial derivatives it is essential to keep the indicated variables constant. If the differentiation is done analytically, then $W_f$ must first be expressed as a function of flux (or flux-linkage) and rotor position only, with current $i$ absent from the expression. Likewise $W'$ must first be expressed as a function of current (or m.m.f.) and rotor position only, with flux (or flux-linkage) absent from the expression. If the differentiation is performed by taking differences or interpolated differences from a look-up table, then the same principle must be observed (Stephenson and Corda 1979).

If magnetic saturation is negligible, then the relationship between flux-linkage and current at the instantaneous rotor position $\theta$ is a straight line whose slope is the instantaneous inductance $L$. Thus

$$\psi = Li$$

and

$$W' = W_f = \frac{1}{2} Li^2 \text{ J}.$$

Therefore

$$T = \frac{1}{2} i^2 \frac{dL}{d\theta} \text{ N m}.$$

If there is magnetic saturation this formula is invalid and the torque should be derived as the derivative of coenergy or field stored energy.

Although saturation plays an important role in determining the characteristics and performance limits of switched reluctance motors, most of the basic control characteristics can be understood from an analysis of the magnetically linear motor, considering only one phase in isolation as in Fig. 7.6(a). Mutual coupling between phases is ignored in this analysis. In practice it is desirable to keep it as small as possible, by making the stator yoke thick enough to prevent cross-saturation effects between phases that share common sections of the magnetic circuit.

As the rotor rotates, the inductance $L$ varies between two extreme values. The maximum $L_a$ occurs when the rotor and stator poles are aligned. The minimum inductance $L_u$ occurs when a rotor interpolar axis is aligned with the stator poles. The variation with rotor position is shown in idealized form in Fig. 7.7, in which the neglect of fringing results in sharply defined 'corners' which coincide with particular positions. If the rotor and stator pole arcs are different, there will be a small 'dwell' at maximum inductance. Likewise if the interpolar arc of the rotor exceeds the stator pole arc, there is a 'dwell' at minimum inductance. The upper and lower corners occur when rotor and stator pole corners are in conjunction, and between these positions the inductance varies more or less linearly as the overlap area varies. If the steel is assumed to be infinitely permeable and fringing is neglected, the inductance can be estimated roughly as

$$L(\theta) = 2N_p^2 P_g + L_u = 2N_p^2 \frac{\mu_0 r_1 l \alpha}{g} + L_u$$

STATIC TORQUE PRODUCTION 161

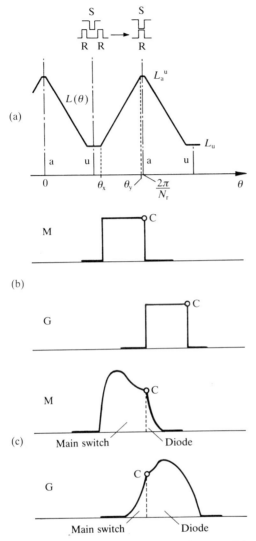

FIG. 7.7. Variation of inductance with rotor position. (a) Idealized inductance variation, ignoring fringing and saturation. (b) Ideal current waveforms for motoring and generating at low and medium speeds. (C = commutation). (c) High-speed current waveforms depart from the ideal, owing to the self-e.m.f. of the motor. Note the phase advance in the turn-on angles relative to those in (b).

where $\alpha$ is the overlap angle between the rotor and stator poles. If the origin of $\theta$ for a particular phase is taken to be the alignment of the previous pair of rotor poles, then $\alpha$ and $\theta$ are related by the following expression throughout the rising-inductance interval:

$$\alpha = \theta - \theta_x; \quad \theta_x < \theta < \theta_y$$

where

$$\theta_x = \frac{2\pi}{N_r} - \frac{\beta_r + \beta_s}{2}; \quad \theta_y = \frac{2\pi}{N_r} - \frac{\beta_r - \beta_s}{2}.$$

The form of Fig. 7.7 reflects the variation of the overlap angle $\alpha$ as the rotor rotates. The unaligned inductance includes the end-turn inductance and a contribution due to leakage flux passing across the stator slots.

The torque is independent of the direction of the current. Its direction depends only on the sign of $dL/d\theta$. When the rotor poles are approaching the aligned position this is positive, and positive (motoring) torque is produced, regardless of the direction of the current. When the rotor poles are leaving the aligned position and approaching the unaligned position, the torque is negative (braking or regenerating), regardless of the direction of the current. Therefore the ideal motoring current waveform is a rectangular pulse that coincides with the rising inductance. Similarly, the ideal braking current waveform is a rectangular pulse that coincides with the falling inductance. The implication is that the current must be switched on and off in synchronism with the rotor position; in other words, the SR motor is a shaft-position-switched machine just like the squarewave brushless d.c. motor.

To produce torque at all rotor positions the entire 360° must be 'covered' by segments of rising inductance from different phases, as shown in Fig. 7.8, and the phase currents must be commutated and sequenced to coincide with the appropriate segments as shown. There is no fundamental reason why the conduction period on each phase should not exceed the step angle, and indeed a small amount of overlap is desirable to minimize torque ripple in the form of notches in the instantaneous torque waveform at the commutation instant. Too much overlap can lead to positive impulses of torque at the commutation angles. While these add to the average torque, they impose transient or vibratory stresses on the shaft, coupling, and load. 'Commutation underlap', on the other hand, is permissible only at speeds high enough so that the rotor inertia (including any load inertia) can maintain rotation through the torque notches. Any underlap at zero speed may result in failure to start if the rotor position happens to fall between the turn-off angle of one phase and the turn-on angle of the next. In the 6:4 motor the minimum conduction angle is 30°, and in the 8:6 motor it is 15°.

Not all SR motors admit the same degree of overlap between phases, because of the geometrical constraints of the pole geometry. With 6:4 three-phase motors the maximum conduction angle is 45°, i.e. 1.5 times the step

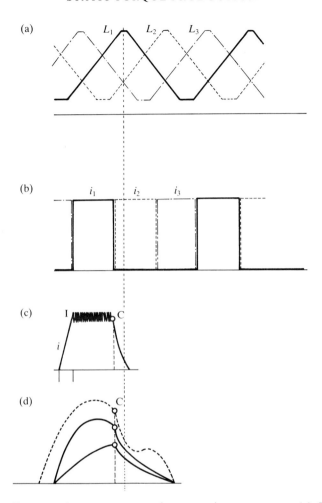

FIG. 7.8. Commutation sequence produces continuous torque. (a) Inductance profiles of three phases. (b) Ideal current waveforms of the three phases at low and medium speeds. With suitable pole geometry these waveforms produce a relatively smooth torque. (c) Practical phase-current waveform at low speed, controlled by chopping or p.w.m. (d) Practical current waveforms at very high speed, limited by the self-e.m.f. of the motor. These waveforms produce significant torque ripple. The lower solid curve is at a higher speed than the upper solid curve, but has the same switching angles and the same supply voltage. This effect causes the torque to fall off at high speed. Advancing the turn-on angle, as shown by the dotted curve, increases the current and the torque, up to the point where the total conduction angle equals the rotor pole pitch.

angle. In the 8:6 four-phase motor, the maximum conduction angle is 30°, i.e. 2.0 times the step angle. The four-phase motor can therefore have more conduction overlap between phases, but it will only be useful if the stator poles can be made sufficiently wide to provide a correspondingly wider angle of rising inductance, without reducing the slot area to the point where copper losses become too high. In other words, the four-phase motor might have less torque ripple but only at the expense of efficiency.

It is also possible to operate the four-phase motor with periods of simultaneous conduction on two phases. While this may increase the torque, it can also increase the torque ripple because of the changes that take place at the commutation angles. Many of these considerations with regard to torque ripple also apply to the squarewave PM brushless d.c. motor (Chapter 4). The practical impossibility of maintaining ideal commutation over a wide speed range is in sharp contrast with the smooth transitions between phases in a.c. sinewave machines.

### 7.3.1 Energy conversion loop

The average torque can be estimated from the 'energy conversion loop' which is the locus described on the flux-linkage/current diagram by the point whose coordinates are $(i, \psi)$ during each step or working stroke. This is shown in Figs 7.9, 7.12, and 7.15. Note the saturation in the aligned magnetization

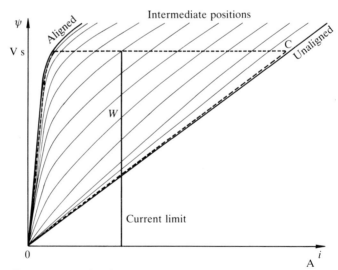

FIG. 7.9. Energy conversion loop showing maximum available energy conversion (dotted trajectory). In practice the current is limited, either thermally or by the converter, as shown by the vertical line. Similarly the maximum flux-linkage is limited by core losses and the tendency towards a spiky current waveform and noisy operation. In small motors only a small fraction of the available conversion energy can be utilized.

# STATIC TORQUE PRODUCTION

curve. Rectangular current pulses can be realized at low speeds if the phase is switched on at the unaligned position and off at the aligned position, and the current is maintained constant by some external means such as p.w.m. regulation or a large external reactor. There is also a 'natural' speed at which the current waveform is flat-topped; this speed and the corresponding current are related by the equation

$$V - Ri = i\omega \frac{dL}{d\theta}$$

(see Section 7.4). Again, this equation is valid only if saturation is negligible.

The electromagnetic energy that is available to be converted into mechanical work is equal to the area $W$. In one revolution each phase conducts as many strokes as there are rotor poles, so that there are $qN_r$ strokes or steps per revolution. The average torque is therefore given by

$$\text{Average torque} = \text{Work per stroke} \times \frac{\text{Number of strokes per revolution}}{2\pi}$$

or

$$T_a = W \frac{qN_r}{2\pi} \text{ N m.}$$

The average electromagnetic power converted is

$$P_e = \omega T_a$$

where $\omega$ is the speed in rad/s. From this must be subtracted the friction and windage and rotor core losses.

The area $W$ shown in Fig. 7.9 represents the maximum energy available for conversion with the flux-linkage limited to the value shown. Obviously the torque per ampere will be maximized if the area between the aligned and unaligned curves is maximized. Ideally this requires

(1) the largest possible unsaturated aligned inductance, implying a small airgap with wide poles;
(2) the smallest possible unaligned inductance, implying a large interpolar arc on the rotor, narrow stator poles, and correspondingly deep slotting on both the stator and the rotor;
(3) the highest possible saturation flux-density.

While the geometry is simple, it is by no means easy to achieve these objectives in design calculations and computer methods are essential to get a good result. The pole geometry that maximizes the 'static' torque per ampere (as derived from Fig. 7.9) can be uniquely determined, following procedures that have long been established for stepper motors. The problem is more complicated for the SR motor because of the unequal rotor and stator slotting, and when the dynamics are taken into account it becomes impossible to establish a single lamination geometry that is optimum for all applications. The main reason for this is that the SR motor is usually applied as a high-speed machine with a wide range of variable speed, and therefore the current waveform is almost never a pure rectangular wave. The current waveform may vary from the one shown in Fig. 7.8(c) at low speeds to the one in Fig. 7.8(d) at high speeds. At low speeds the current is forced to have an approximately rectangular shape by chopping or p.w.m. in the controller. At high speeds there is no chopping and the current waveform takes up a natural shape determined by the speed, the turn-on and turn-off angles, the applied voltage, and the rate of change of inductance.

The static energy-conversion diagram is a useful first step in design. It requires the calculation of only the aligned and unaligned magnetization curves, which are both amenable to finite-element analysis as shown in Fig. 7.10. It provides an upper bound on the torque capability, which can be approached at low speeds.

If the vertical line representing the maximum permissible current is swept from left to right, the area $W$ initially increases with the square of the current, but as saturation sets in it becomes more nearly linear. The torque per ampere thus becomes more nearly constant as the current increases.

In small motors only a small fraction of the available energy can be converted because of the thermal current limit. An increase in scale naturally permits more of the available energy to be converted. The same increase could be obtained with more intense cooling or during intermittent operation. However, operation with an extreme conversion loop may be inefficient and noisy, with a poor power factor and peaky currents.

A rough estimate of the maximum attainable torque per unit rotor volume can be derived from an idealized triangular area approximating the dotted trajectory in Fig. 7.9. Following methods of Harris (1975) the result is

$$\frac{T}{V} = \frac{N_r B_s^2 (\lambda - 1) q}{2\pi^2 \mu_0} \frac{\beta g}{r_1} \text{ N m/m}^3$$

where $B_s$ is the flux-density in the stator poles at the maximum flux-linkage $\psi_s$ in the aligned position; $\lambda$ is the aligned/unaligned unsaturated inductance ratio; $\beta$ is the pole arc (assumed equal for stator and rotor); and $g$ is the airgap. For a three-phase 6:4 motor having a pole arc of 30° and an airgap of 0.25 mm, at a rotor radius $r_1$ of 25 mm, it should be possible to achieve $\lambda = 10$ and $B_s = 1.6$ T, giving a specific torque of 60 kN m/m$^3$ from the extreme trajectory. With very small motors (say, less than 100 W) the fraction of this theoretically available torque that can be achieved continuously with quiet operation and acceptable losses is only of the order of 5 per cent. This figure improves rapidly with scale so that at 5 kW it should be possible to achieve up to 25 per cent or 15 kN m/m$^3$, and in highly-rated machines with special cooling perhaps double this; transient ratings may be still higher.

# STATIC TORQUE PRODUCTION

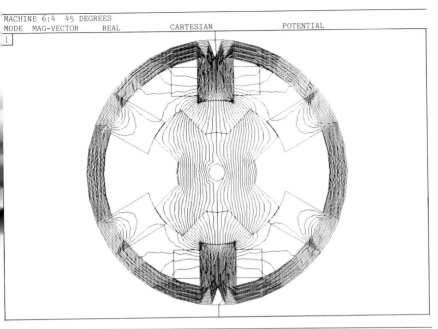

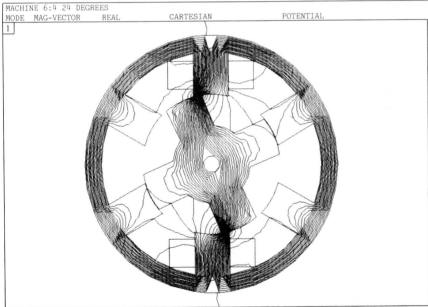

FIG. 7.10. Finite-element flux-plot for SR motor. Courtesy Lucas Engineering & Systems Ltd.

## 7.4 Partition of energy and the effects of saturation

The shape of the energy conversion loop, and its area $W$, depend on the variation of current with rotor angle, and this in turn depends on the control parameters and the speed, as well as on the motor design. The SR motor is unlike most other motors in that the current waveform may vary widely over the operating range. To determine both the current and the torque at speed it is necessary to simulate the operation of the motor (and converter) for at least one stroke. This means solving the terminal voltage equation as a function of time, and since this is a differential equation a time-stepping method such as Euler's method or the Runge-Kutta procedure is required. In the case of the ideal machine with no fringing or magnetic saturation, represented by the inductance variation shown in Fig. 7.7, an analytical solution is possible, as elegantly presented by Ray and Davis (1979), in which many of the basic characteristics of the SR drive are identified and illustrated. Here, however, the solution by numerical techniques will be outlined: such methods are necessary in practice to deal with saturation and fringing effects.

The terminal voltage equation for one phase is

$$v = Ri + \frac{d\psi}{dt}.$$

Suppose that the flux-linkage $\psi$ is a function of both current $i$ and rotor angle $\theta$:

$$\psi = \psi(i, \theta).$$

Then

$$\frac{d\psi}{dt} = \frac{\partial \psi}{\partial i}\frac{di}{dt} + \omega \frac{\partial \psi}{\partial \theta} = L\frac{di}{dt} + e$$

where $L$ is the incremental inductance (the slope of the magnetization curve) and $e$ is a 'back-e.m.f.'. This equation is quite general and shows that from the terminals the SR motor appears to have an equivalent circuit that comprises, in each phase, a resistance, an incremental inductance, and a back-e.m.f. In a general way this is similar to other motors. But the back-e.m.f. is different in that it depends on the phase current, being ideally proportional to it. The back-e.m.f. also varies with rotor position, and in general it cannot be regarded as the only term in the voltage equation that contributes to the torque production. The term 'self-e.m.f.' will be used instead, because this carries the implication that the product $ei$ includes an energy-storage component as well as an energy-conversion component. During the interval of rising inductance, $e$ varies strongly with current and weakly with rotor position, but $L$ varies strongly with rotor position and weakly with current.

Apart from losses, the electrical energy supplied at the terminals during a

## ENERGY PARTITION AND SATURATION

small rotation is partitioned between stored magnetic field energy and mechanical work. Ideally all the energy would be converted to mechanical work, but this cannot be achieved in practice. The proportion in which the energy divides between magnetic field energy and mechanical work depends on the shape of the magnetization ($\psi$, $i$) curves in the neighbourhood of the particular rotor position.

If there is no saturation the incremental inductance is the total inductance at the particular rotor angle, and this is equal to the ratio of flux-linkage to current. In this case

$$\frac{d\psi}{dt} = L\frac{di}{dt} + i\omega\frac{dL}{d\theta}.$$

The first term has the appearance of inductive voltage drop across a fixed inductance, while the second term is the self-e.m.f. proportional to current, speed, and rate of change of inductance with rotor angle. If the current is flat-topped, then during the flat-top period the self-e.m.f. is constant and the first term is zero. This defines the 'natural' speed at which a flat-topped current waveform is achieved without chopping, as discussed in Section 7.3. On the other hand, if the inductance is constant (as for example around the unaligned position), the self-e.m.f. is zero and the first term absorbs all the applied voltage. The equivalent circuit can change from being mainly an inductance to mainly an e.m.f., depending on the rotor angle and the current waveform.

If the above equation is multiplied by $i$ the left-hand side represents the electrical power supplied (after resistive losses have been subtracted). The rate of change of stored magnetic energy is

$$\frac{d}{dt}[\tfrac{1}{2} Li^2] = iL\frac{di}{dt} + \tfrac{1}{2}i^2\frac{dL}{d\theta}\omega.$$

Subtracting this equation from the previous one multiplied by $i$, the electromechanical energy conversion is

$$P_m = T\omega = \tfrac{1}{2}i^2\frac{dL}{d\theta}\omega$$

which gives the 'linear' expression for instantaneous torque as before. Of the electrical power supplied, however, this represents less than half. The remainder is the rate of magnetic energy storage, which exceeds the electromechanical conversion by the term

$$iL\frac{di}{dt} + \tfrac{1}{2}i^2\frac{dL}{d\theta}\omega.$$

The most 'effective' use of the energy supplied is when the current is maintained constant (during a period of rising inductance), and even then the highest level of 'effectiveness' is only 50 per cent. This is illustrated in

Fig. 7.11(a), which shows the energy exchanges over a small rotation $\Delta\theta$. The triangular area $\Delta W_m$ representing the mechanical energy conversion is one-half the rectangular area $\Delta W_e$ representing the electrical energy supplied, since it has the same base ($\Delta\psi$) and the same height $i$, the current $i$ being constant through the small rotation.

Note that the energy stored in the magnetic field is not necessarily dissipated. With the appropriate converter circuit it can be recovered to the supply at the end of the period of rising inductance. This is why the term 'effectiveness' is used, not 'efficiency'. The consequence of a low 'effectiveness' is to increase the volt-ampere rating of the converter for a given power conversion in the motor. 'Effectiveness' is therefore akin to power factor in a.c. machines, and it can be defined more precisely in terms of the energy ratio defined below.

The partition of input energy into mechanical work and stored field energy is improved if the motor saturates. This has been discussed very clearly by Byrne (1972). If, at a given rotor position, the magnetization curve is saturated as in Fig. 7.11(b), then the area representing the mechanical work can exceed half the area of the 'supply rectangle'. An extreme case is shown in Fig. 7.11(c), with very sharp saturation occurring at a low current level and extremely high inductance below this level. In this case practically all the energy supplied is converted to mechanical work and very little is stored in the magnetic field. The effectiveness theoretically approaches 100 per cent. But such curves could not be realised with practical electrical steels, and they would require very small or zero airgap clearances.

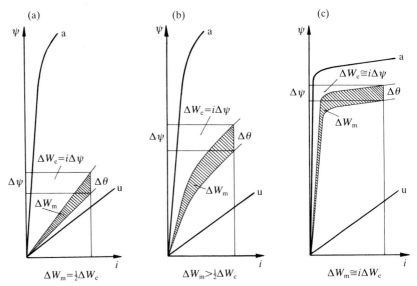

FIG. 7.11. Partition of input electrical energy. (a) Linear (no saturation). (b) Typical practical case. (c) Idealized case with extreme saturation.

# ENERGY PARTITION AND SATURATION

This argument has been developed for incremental exchanges of energy taking place over very small rotations. Similar principles apply to the total exchanges taking place in a complete working stroke, see Fig. 7.12. The total electromechanical energy converted is $W$, as before. The energy returned by the motor to the external circuit is $R$; assume that all of this is returned to the supply following commutation at C. Note that $R$ is smaller than the actual stored magnetic energy at C, a proportion of which is converted to mechanical work during the freewheeling period. The apparent power supplied by the external circuit is $W+R$. This is proportional to the product of the voltage-time integral and either the peak or the mean current, and is therefore the primary determinant of the converter rating. The energy ratio is defined as

$$\frac{W}{W+R}.$$

It is shown in Miller (1985) that in a magnetically linear motor

$$\frac{W}{W+R} = \frac{\lambda_u - 1}{2\lambda_u - 1}$$

where $\lambda_u$ is the inductance ratio between the aligned and unaligned positions. With a ratio of 6, the energy ratio is 0.455. It improves with the inductance ratio, but can never exceed 0.5.

In the saturating motor the shape of the saturating magnetization curves helps to reduce the ratio between $R$ and $W$, and the energy ratio is greater. In

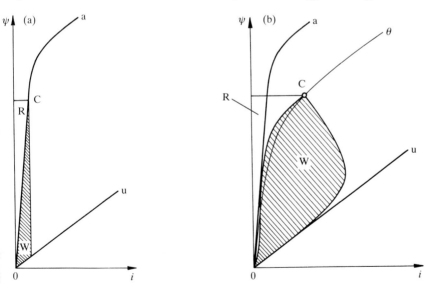

FIG. 7.12. Energy ratio and energy exchanges during one complete working stroke. (a) Linear case. (b) Typical practical case. W = energy converted into mechanical work. R = energy returned to the d.c. supply via freewheel diodes.

practical machines the maximum value is still far from the ideal of unity (which would require zero unaligned inductance in addition to the special saturation characteristics mentioned earlier for the aligned curve). Values of 0.6–0.7 are not difficult to achieve, however, and this turns out to be sufficient to keep the converter volt-amperes in the same range as for inverter-fed induction motors (Harris *et al.*, 1986; Miller, 1985). However, this is only possible because of the increased switching frequency (the vernier effect). For the same number of rotor poles the SR motor has roughly half the electromagnetic 'effectiveness' of a good a.c. motor, but it traverses the energy conversion loop twice as often, restoring the converter volt-amperes to about the same level as in a.c. drives, but only at the expense of higher switching frequency (i.e., commutation frequency, not chopping frequency).

The concept of energy ratio can be applied also to a.c. machines (on a per-phase basis); when this is done it is found that even for quite low values of energy ratio the terminal power factor remains high and the converter volt-amperes remain low. For the motors discussed in Chapter 6, the energy ratio is naturally much lower for the synchronous reluctance motor than for the PM motors.

In the literature there has been much discussion of the 'torque doubling' effect of saturation. While saturation improves the energy ratio, it decreases the conversion area $W$ per unit of motor volume. The decrease in converter kVA/kW outweighs the decrease in power per unit volume, and for this reason saturation is desirable up to the point where the increased core losses associated with it limit the power density. It is not the case that high-speed reluctance machines can produce 'double the torque' and it is equally untrue that they require 'double the kVA'. When all the complexities of design are taken fully into account, their performance in terms of these parameters is very roughly on a par with that of induction motors.

## 7.5 Dynamic torque production

Under normal operating conditions at speed, the energy exchanges, both incremental and total, can be determined by integrating the voltage equation and developing the conversion loop in the $\psi, i$ diagram. The necessary time-stepping procedure was developed by Stephenson and Corda (1979) and only the outline of their method is described here.

The voltage equation is integrated in the form

$$\psi = \int (v - Ri) \, dt$$

through one time-step, giving a new value of $\psi$. If the speed is assumed constant, the integration can be done with respect to rotor angle $\theta$. Otherwise the rotor angle must be determined by a simultaneous integration of the mechanical equations of motion, as is normal in such simulations. At the end of the time-step, $\theta$ and $\psi$ are both known, and the current $i$ can be determined from the magnetization curve for that rotor angle. To minimize this computation Stephenson and Corda used a set of polynomials to represent the magnetization curves at a number of rotor angles between the aligned and unaligned positions, and then applied an interpolation procedure at the end of each time-step to determine the current from the flux-linkage at the particular rotor position. The instantaneous torque can be determined from the difference-approximation to the partial derivative of coenergy at constant current, by a second interpolation procedure that uses stored field energies precalculated at discrete current levels at each of a number of rotor angles. By this method the waveforms of instantaneous phase current and torque can be developed from the integration. An example is shown in Figs 7.13(a–c), 7.14(a–c), and 7.15(a–c).

## 7.6 Converter circuits

The torque is independent of the direction of the phase current, which can therefore be unidirectional. This permits the use of unipolar controller circuits with a number of advantages over the corresponding circuits for a.c. or PM brushless motors, which require alternating current. Although the SR motor could be operated with alternating (but non-sinusoidal) current, unidirectional (d.c.) current has the added advantage of reducing hysteresis losses.

Figure 7.16(a) shows a circuit well suited for use with transistors (bipolar, field-effect, or insulated-gate). The phases are independent, and in this respect the SR controller differs from the a.c. inverter, in which the motor windings are connected between the midpoints of adjacent inverter phaselegs. The winding is in series with both switches, providing valuable protection against faults. In the a.c. inverter the upper and lower phaseleg switches must be prevented from switching on simultaneously and shorting the d.c. supply; this is possible only by means of additional control circuitry, which is unnecessary in the SR controller.

The upper and lower phaseleg switches are switched on together at the start of each conduction period or working stroke. At the commutation point (C in Fig. 7.7 and 7.8) they are both switched off. During the conduction period either or both of them may be chopped according to some control strategy, such as maintaining the current within a prescribed 'hysteresis band'. This mode of operation is necessary at low speeds when the self-e.m.f. of the motor is much smaller than the supply voltage. At high speeds both transistors remain on throughout the conduction period and the current waveform adopts a 'natural' shape depending on the speed and torque. It is convenient in the logic design to use one transistor primarily for 'commutation' and the other for regulation or chopping. At the end of the conduction period when both switches are turned off, any stored magnetic energy that has not been

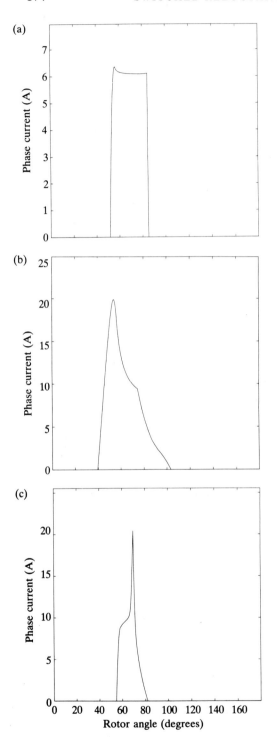

FIG. 7.13. Current waveform produced by Stephenson and Corda's method. (a) Low and medium speeds. The natural flat-toppped waveform arises when the self-e.m.f. equals the applied voltage. (b) Very high speed. The self-e.m.f. exceeds the applied voltage once the poles begin to overlap. Note the phase advance; the turn-on angle is at 40°, i.e. 5° before the unaligned position. Commutation is also advanced relative to that in (a). (c) Low and medium speeds, but with a higher applied voltage than in (a). The motor is being driven harder to convert more of the available energy than in (a). The conduction angle is narrower than in (a), but because of the high driving voltage the stator core saturates, causing a peak to appear in the current waveform. To avoid this, the voltage must be reduced, or the commutation angle advanced, or the stator yoke must be made thicker.

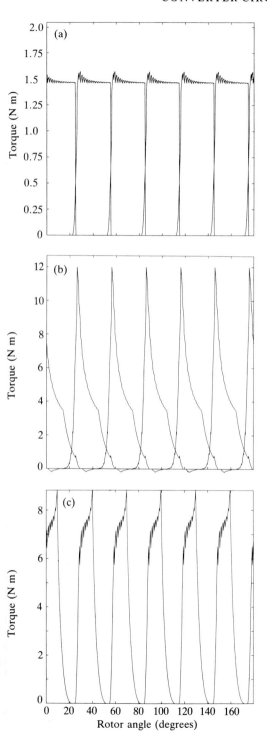

FIG. 7.14. Torque waveform of the three phases corresponding to Fig. 7.13. The high-frequency ripple effect, especially in (a) and (b), is due to the numerical approximation of the magnetization curves and does not appear in the current waveform. (a) Low and medium speeds. (b) High speed. (c) Low and medium speed with increased drive voltage.

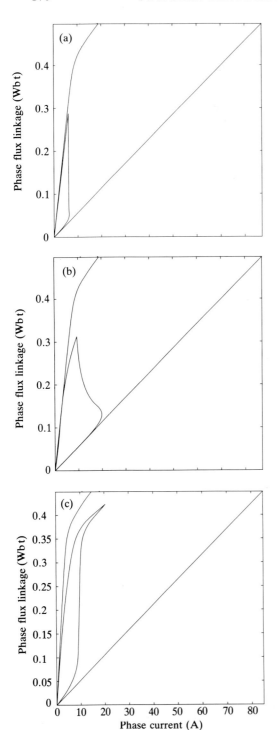

FIG. 7.15. Energy conversion loops corresponding to Figs. 7.13 and 7.14. (a) Low and medium speeds. (b) High speed. (c) Low and medium speed with increased drive voltage. The converted energy (and average torque) is much higher in (b), showing how the increase in conduction angle overcomes the effects of increasing self-e.m.f. at high speeds. This is an example of the 'programmability' of the speed/torque characteristics.

# CONVERTER CIRCUITS 177

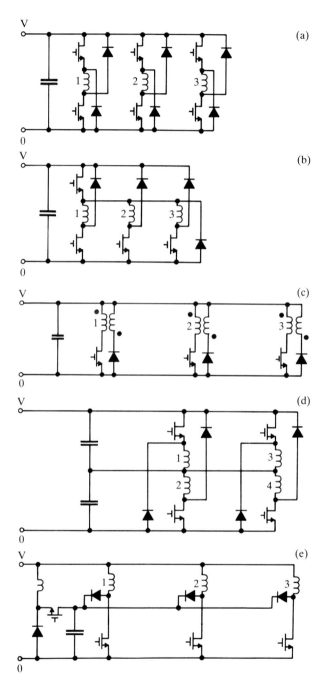

7.16. Converter circuits for three-phase SR motor. (a) Two-transistor/phase circuit. (b) $n+1$ transistors for $n$-phase motor. (c) Bifilar windings. (d) Split-link circuit used with even phase-number. (e) C-dump circuit.

converted to mechanical work is returned to the supply by the current freewheeling through the diodes. Note that when they become forward-biased, the diodes connect the negative of the supply voltage across the winding to reduce its flux-linkage quickly to zero.

The inductance varies with rotor position. Therefore if fixed-frequency chopping is used, the current ripple varies. If hysteresis-type current regulation is used, the chopping frequency varies as the poles approach alignment. Figures 7.17–7.19 show operation in the chopping mode. Chopping frequencies above 10 kHz are usually desirable, as in other types of drive, to minimize acoustic noise. In larger drives (say, above 20 kW) it becomes progressively more difficult to chop at such a high frequency because of the limited switching speed and losses of larger devices (such as GTOs). This makes it more difficult to achieve quiet operation in larger motors, particularly at low speeds.

In small drives it is often acceptable to use p.w.m. control over the entire speed range. This is also the usual approach with surface-magnet PM brushless motors. In such cases the SR controller circuit can profitably be reduced to the circuit of Fig. 7.16(b), in which the chopping is performed by one transistor in common for all the phases. The lower transistors commutate the chopped voltage to the phases in proper sequence, under the control of the shaft position sensor and gating logic. This circuit requires only $n+1$ transistors and $n+1$ diodes for a motor with $n$ phases. A three-phase motor thus requires only four transistors and four diodes. There is practically no loss of functionality with this circuit relative to the full circuit having $2n$ transistors

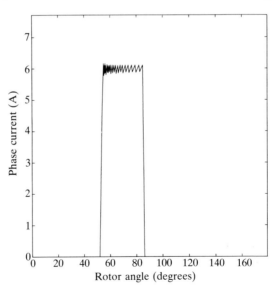

FIG. 7.17. Current waveform in chopping mode at low speed. Note the reduction in chopping frequency as the phase inductance increases with increasing overlap.

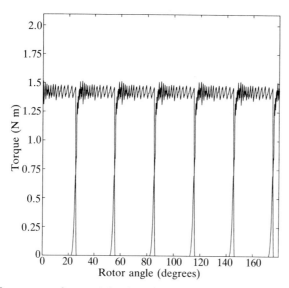

FIG. 7.18. Torque waveforms of the three phases, corresponding to Fig. 7.17.

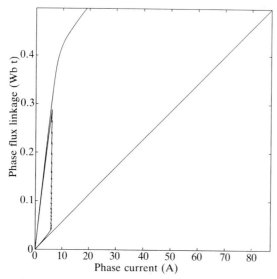

FIG. 7.19. Energy conversion loop corresponding to Fig. 7.17.

in Fig. 7.16(a), and indeed at low speeds it tends to run more smoothly. Its main limitation is that at very high speeds the phases cannot be 'de-fluxed' or de-energized fast enough through the diodes, because the control transistor keeps switching on, with a long duty cycle. If there is still a significant diode current by the time the rotor reaches the aligned position, it may start to increase as the self-e.m.f. becomes negative, and braking torque is produced. As the chopping duty cycle and/or the speed increase further, the net torque decreases rapidly and the losses increase. These problems only arise when the speed range is very wide, typically more than 20:1.

Many other circuits have been developed in attempts to reduce the number of switches all the way down to $n$ and take full advantage of unipolar operation (see Bass et al., 1987 for a review). The split-link circuit in Fig. 7.16(d) has been successfully used by Tasc Drives Ltd. with GTO thyristors in a range of highly efficient drives from 4–80 kW. In other cases it seems that when the device count is reduced to one per phase, there is a penalty in the form of extra passive components or control limitations. The bifilar winding in Fig. 7.16(c) suffers from double the number of connections, a poor utilization of copper, and voltage spikes due to imperfect coupling between the bifilar windings. In Fig. 7.16(e) the device count is reduced to $n$ plus one additional device to bleed the stored energy from the dump capacitor C back to the supply via the step-down chopper circuit. The mean capacitor voltage is maintained well above the supply rail to permit rapid de-fluxing after commutation. A control failure in the energy-recovery circuit would result in the rapid build-up of charge on the dump capacitor, and if protective measures were not taken the entire converter could fail from overvoltage. With more than three phases, substantial reductions in the number of devices per phase appear to be achieved by the methods of Pollock and Williams (1987).

## 7.7 Control: current regulation, commutation

For motoring operation the pulses of phase current must coincide with a period of increasing inductance, i.e. when a pair of rotor poles is approaching alignment with the stator poles of the excited phase. The timing and dwell of the current pulse determine the torque, the efficiency, and other parameters. In d.c. and brushless d.c. motors the torque per ampere is more or less constant, but in the SR motor no such simple relationship emerges naturally. With fixed firing angles, there is a monotonic relationship between average torque and r.m.s. phase current, but in general it is not very linear. This may present some complications in feedback-controlled systems although it does not prevent the SR motor from achieving 'near-servo quality' dynamic performance, particularly in respect of speed range, torque/inertia, and reversing capability.

The general structure of a simple control scheme is much the same as that of the brushless d.c. drive (Fig. 7.14). More complex controls are required for higher-power drives, particularly where a wide speed range is required at constant power, and microprocessor controls have been developed and used very effectively, (Chappell et al. 1984; Bose et al. 1986). Because the characteristics of the SR drive are essentially controlled by the phasing of switching instants relative to the rotor position, digital control is not only very natural but can be implemented extremely effectively with flexibility or 'programmability' of the characteristics and with reliable, repeatable results.

It is characteristic of good operating conditions that the conversion loop fits snugly in the space between the unaligned and aligned magnetization curves, as in Figs 7.15 and 7.19. Figure 7.15(b) corresponds to high-speed operation where the peak current is limited by the self-e.m.f. of the phase winding. A smooth current waveform is obtained with a peak/r.m.s. ratio similar to that of a half sinewave.

At low speeds the self-e.m.f. of the winding is small and the current must be limited by chopping or p.w.m. of the applied voltage. The regulating strategy employed has a marked effect on the performance and the operating characteristics. Figure 7.17 shows a current waveform controlled by a 'hysteresis-type' current-regulator that maintains a more or less constant current throughout the conduction period in each phase. Figure 7.20(a) shows schematically the method of control. As the current reference increases, the torque increases. At low currents the torque is roughly proportional to current squared, but at higher currents it becomes more nearly linear. At very high currents saturation decreases the torque per ampere again. This type of control produces a constant-torque type of characteristic as indicated in Fig. 7.21. With loads whose torque increases monotonically with speed, such as fans and blowers, speed adjustment is possible without tachometer feedback, but in general feedback is needed to provide accurate speed control. In some cases the pulse train from the shaft position sensor may be used for speed feedback, but only at relatively high speeds. At low speeds a larger number of pulses per revolution is necessary, and this can be generated by an optical encoder or resolver, or alternatively by phase-locking a high-frequency oscillator to the pulses of the commutation sensor (Bose 1986). Systems with resolver-feedback or high-resolution optical encoders can work right down to zero speed. The 'hysteresis-type' current regulator may require current transducers of wide bandwidth, but the SR drive has the advantage that they

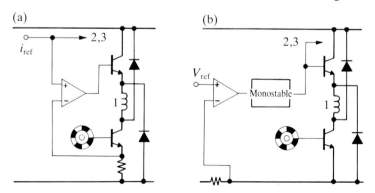

FIG. 7.20. Schematic of current-regulator for one phase. (a) Hysteresis-type. (b) Voltage-p.w.m. type (duty-cycle control).

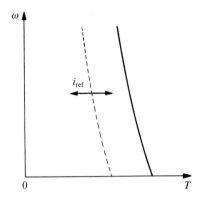

FIG. 7.21. Constant-torque characteristic obtained with regulator of Fig. 7.20(a).

can be grounded at one end, with the other connected to the negative terminal of the lower phaseleg switch. Shunts or Hall-effect sensors can be used, or alternatively, 'Sensefets' with in-built current sensing. Much of the published literature on SR drives describes this form of control.

Figure 7.20(b) shows an alternative regulator using fixed-frequency p.w.m. of the voltage with variable duty-cycle. The current waveform is similar to that shown in Fig. 7.13, except that after commutation the current decays through the diodes somewhat more rapidly because the reverse voltage applied is effectively $1/d$ times the forward voltage applied before commutation. ($d$ = duty cycle). The torque and energy-conversion loop are similar to Figs 7.14 and 7.15. The duty-cycle (or 'off-time') of the p.w.m. can be varied by a simple monostable circuit. This form of control is similar to armature-voltage control in a d.c. motor.

Current feedback can be added to the circuit of Fig. 7.20(b) to provide a signal which, when subtracted from the voltage reference, modulates the duty cycle of the p.w.m. and 'compounds' the torque-speed characteristic. It is possible in this way to achieve under-compounding, over-compounding, or flat compounding just as in a d.c. motor with a wound field. For many applications the speed regulation obtained by this simple scheme will be adequate. For precision speed control, normal speed feedback can be added. The current feedback can also be used for thermal overcurrent sensing.

A desirable feature of both the 'hysteresis-type' current-regulator and the voltage p.w.m. regulator is that the current waveform tends to retain much the same shape over a wide speed range.

When the p.w.m. duty cycle reaches 100 per cent the motor speed can be increased by increasing the dwell (the conduction period), the advance of the current-pulse relative to the rotor position, or both. These increases eventually reach maximum practical values, after which the torque becomes inversely proportional to speed squared, but they can typically double the speed range at constant torque. The speed range over which constant power can be maintained is also quite wide, and very high maximum speeds can be obtained, as in the synchronous reluctance motor and induction motor, because there is not the limitation imposed by fixed excitation as in PM motors.

### 7.7.1 Torque/speed characteristic

The generic form of the torque/speed capability curve is shown in Fig. 7.22. For speeds below $\omega_b$ the torque is limited by the motor current (or the controller current, whichever is less). Up to the 'base speed' $\omega_b$ it is possible, by means of the regulators in Fig. 7.20, to get any value of current into the motor, up to the maximum. The precise value of current at a given operating point depends on the load characteristics, the speed, and the regulator and control strategy. In the speed range below $\omega_b$ the firing angles can be chosen to optimize efficiency or minimize torque ripple. If the load never needs to operate at high speeds above $\omega_b$, it will usually be possible to design the pole geometry to optimize these parameters without regard to the efficiency at high speeds, and this provides considerable design freedom to obtain smooth torque and simplify the control.

The 'corner point' or base speed $\omega_b$ is the highest speed at which maximum current can be supplied at rated voltage, with fixed firing angles. If these angles are still kept fixed, the maximum torque at rated voltage decreases with speed squared. However, if the conduction angle is increased (mainly by advancing the turn-on angle) there is a considerable speed range over which maximum current can still be forced into the motor, and this sustains the torque at a level high enough to maintain a constant-power characteristic, even though the

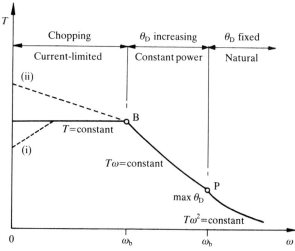

FIG. 7.22. General torque/speed characteristic of switched reluctance motor.

core losses and windage losses increase quite rapidly with speed. This is shown in Fig. 7.22 between points B and P. The angle $\theta_D$ is the 'dwell' or conduction angle of the main switching device in each phase. It should generally be possible to maintain constant power up to 2–3 times base speed.

The increase in conduction angle may be limited by the need to avoid continuous conduction, which occurs when the conduction angle exceeds half the rotor pole-pitch. It may be limited to lower values by the core loss or other factors. At P the increase in $\theta_D$ is halted and higher speeds can now only be achieved with the natural characteristic, i.e. torque decreasing with speed squared.

At very low speeds the torque/speed capability curve may deviate from the flat-torque characteristic. If the chopping frequency is limited (as with GTO thyristors, for example), or if the bandwidth of the current regulator is limited, it may be difficult to limit the peak current without the help of the self-e.m.f. of the motor, and the current reference may have to be reduced. This is shown in curve (i) in Fig. 7.22. On the other hand, if this is not a problem, the very low windage and core losses may permit the copper losses to be increased, so that with higher current a higher torque is obtained, as shown in curve (ii). Under intermittent conditions, of course, very much higher torques can be obtained in any part of the speed range up to base speed. In Fig. 7.23 this can be seen by extrapolating the constant-duty-cycle curves above the maximum current locus.

It is important to note that the current which limits the torque below base speed is the motor current (or converter output current). The d.c. supply

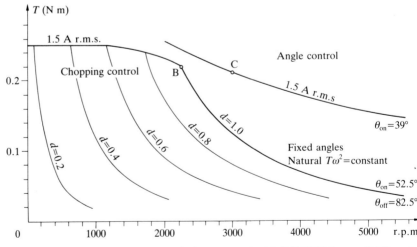

FIG. 7.23. Torque/speed characteristic computed with PC-SRD for small SR motor designed for comparison with PM motors in Chapter 6.

current increases from a small value near zero speed to a maximum value at base speed. Basically this is because the power increases in proportion to the speed as long as the torque is constant. With fixed d.c. supply voltage at the input to the converter, the d.c. supply current is then proportional to the speed. If the torque is less than maximum, of course the d.c. supply current is also smaller.

Figure 7.23 shows the computed torque/speed characteristics of a small motor designed to compare with the PM and synchronous reluctance motors in Chapter 6. The stator diameter and length are identical to those of the PM motors, and the same weight of copper is used. The computation was performed using PC-SRD, a commercially licenced CAD package for SR drives, based on the IBM PC and developed at Glasgow University. The essentially constant-torque characteristic is maintained up to point B at 2250 r.p.m., limited by maximum motor current which corresponds to a winding current-density of 4.8 A/mm$^2$, exactly the same as the a.c. motors in Chapter 6. At speeds above the base speed the natural characteristic is shown at rated voltage, with torque decreasing roughly as speed squared. In the computation, the windage loss was arbitrarily set to zero to make the results a little more general. The effect of windage loss is to increase the torque roll-off particularly above 4000 r.p.m.

The natural characteristics for different fixed values of the chopping duty-cycle $d$ are shown in Fig. 7.23. This parameter has much the same effect as varying the d.c. supply voltage. Also shown in Fig. 7.23 is the effect of conduction angle control. With the chopper saturated, i.e. $d=1$, the applied voltage remains at its rated value and as the speed is increased, the maximum torque is sustained by advancing the turn-on angle with fixed commutation angle. The decrease in the torque is mainly due to the fact that more of the current is being conducted when the rotor is in a position of low $dL/d\theta$, but core losses also contribute to the decrease. If windage losses are included, the net characteristic is still constant-power up to more than 5000 r.p.m.

Point C in Fig. 7.23, at 3000 r.p.m., is comparable with the operating points of the a.c. motors in rows 1, 5, 6, and 10 of Table 6.2. However, the SR motor has lower core losses than the PM motors, so much so that operation at point C can be regarded as thermally continuous, whereas the operating points for the PM motors are intermittent. Figure 7.24 shows the graphical and printed output from PC-SRD for this point.

The variation of torque with current, and the variation of speed with duty-cycle, are in general less linear than in d.c. or brushless d.c. squarewave motors, but they are monotonic, well-behaved functions that are not difficult to accommodate in a controller, and are certainly no more complex than the control laws of a.c. induction or PM synchronous motors. The SR motor has all of the 'programmability' of the brushless d.c. PM motor, but with the added flexibility that comes with angle control, which can provide a wider range of operating speeds for a given converter rating.

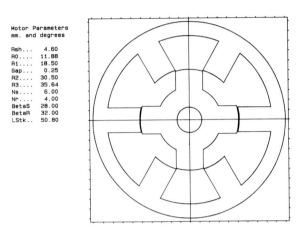
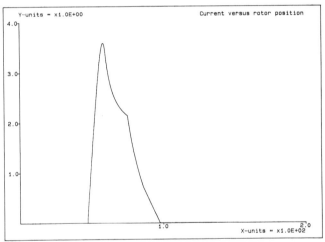
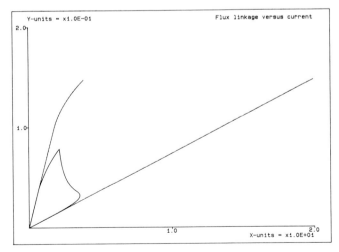

Fig. 7.24. PC-SRD output corresponding to point C in Fig. 7.23.

# CONTROL: CURRENT REGULATION, COMMUTATION

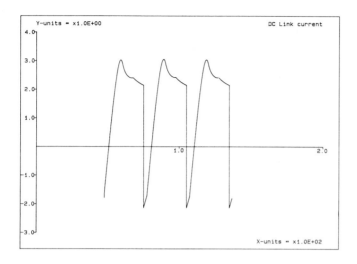

```
18th APR 1988 15:20    PC-SRD Glasgow University 1988            Ver 2.0
-----------------------------------------------------------------------------
 PC-SRD. Sample
-----------------------------------------------------------DIMENSIONS ( mm  )
    Rotor  R0      11.879
           R1      18.500      4.000  Poles   Arc   32.000° Gap         0.250
    Stator R2      30.500      6.000  Poles   Arc   28.000° Stack Lgth 50.800
           R3      35.637
    Stckng fact.    0.970          Steel : Losil 500/50
----------------------------------------------------------------WINDING DATA
   Wire Dia.   = 0.629 mm    3 Phases     Turns/Pole       =     110
   C/S. Areas  :
     1 Wire    = 3.108E-01 Sq.mm          Resistance/Phase =     2.163 Ohm
     Slot Cu   = 3.418E+01 Sq.mm          Temp.                 75.000 °C
   Slot fill    3.500E-01
   M.L.T       = 1.467E+02 mm             Inductance/Phase = 5.556E-02 H Al
   Lgth o/ends = 7.370E+01 mm                              = 7.379E-03 H Un
   Copper wt.  =    0.267 kg                    Ratio     =     7.529
----------------------------------------------------------------CONTROL DATA
   Voltage               60.000           Current Regulator setting
   R.P.M               3000.000                                 1000.000 A
   Turn-on angle         47.400 deg.      Duty Cycle              1.000
   Turn-off angle        75.000 deg.      Transistor RQ           0.000 Ohm
   Dwell angle    =      27.600 deg.      Transistor VQ           2.000 V
   Stroke angle   =      30.000 deg.      Diode      VD           0.600 V
   O/lap starts   =      30.000 deg. BTC  Phase freq.    =      200.000 Hz
-----------------------------------------------------------------PERFORMANCE
   Torque         =   2.113E-01 N-m       Efficiency        =    76.358 %
   Shaft Power    =   6.639E+01 W         kVA/kW(pk)        =    14.868
                                          kVA/kW(rms)       =     5.850
   Losses: Copper =   1.464E+01 W
           Iron   =   5.912E+00 W
           Windage=   0.000E+00 W         Deg. C / W        =     3.000
           Total  =   2.056E+01 W         Temp. rise        =    61.668 °C
   CURRENTS =                    PEAK              MEAN              R.M.S
   Winding                      3.591             0.954              1.502
   Transistor                   3.591             0.745              1.413
   Diode                        2.143             0.209              0.510
   DC Link (Supply)                               1.609
   DC Link (Capacitor)                                                1.517
   RMS Current Density   =  3118.934 A/SQ.in. =    4.834 A/SQ.mm.
-----------------------------------------------------------------------------


                                                      ----SUPPLEMENTARY OUTPUT
   WEIGHTS: Copper =        0.267 kg         Inertia   = 2.589E-05 kg-m²
            Iron   =        0.896
            Total  =        1.163
   Resistivity     = 8.197E-07 Ohm-m         Temp. fact =     1.216
   CPU             =        3.741            ETF        =     1.277
   PSlot           =        1.496            PRS        =     2.245

   IRON LOSSES          Eddy current                         Hysteresis
   Rotor yoke    =        0.444 W                      =       0.243 W
   Rotor poles   =        0.329                        =       0.158
   Stator yoke   =        2.530                        =       1.224
   Stator poles  =        0.671                        =       0.313
   Total         =        3.974                        =       1.938
   Sigma         =        0.280 psi

   End of design
```

### 7.7.2 Shaft position sensing

The commutation requirement of the SR motor is very similar to that of a PM brushless motor. The shaft position sensor and decoding logic are very similar and in some cases it is theoretically possible to use the same shaft position sensor and even the same integrated circuit to decode the position signals and control the p.w.m. as well.

Much has been made of the undesirability of the shaft position sensor, because of the associated cost, space requirement, and possible extra source of potential failures. However, the sensing requirement is no greater and no less than that of the PM brushless motor, and reliable methods are well established. In position servos or speed servos, resolvers or optical encoders may be used to perform all the functions of providing commutation signals, speed feedback, and position feedback.

Operation without the shaft sensor is possible and several schemes have been reported (e.g. Lumsdaine *et al.* 1986, Bass 1987). But to achieve the performance possible with even a simple shaft sensor (such as a slotted disk or a Hall-effect device), considerable extra complexity is necessary in the controller, particularly if good starting and running performance is to be achieved with a wide range of load torques and inertias. For rapid acceleration and/or deceleration cycles, or for position control, there is a long way to go before the sensor can be eliminated. Much the same is true of the PM brushless motor and the induction motor.

When the SR motor is operated in the 'open-loop' mode, like a stepper motor in its slewing range, the speed is fixed by the reference frequency in the controller as long as the motor maintains 'step integrity' (stays in synchronism). Like an a.c. synchronous motor, the SR motor then has a truly constant-speed characteristic. This type of control would be ideal for many applications but it suffers from two difficulties: one is to ensure that synchronism is maintained even though the load torque (and inertia) may vary; the other is to ensure reliable starting. Because of the large step angle and a lower torque/inertia ratio, the SR motor usually does not have the reliable 'starting rate' of the stepper motor, and some form of 'inductance sensing' or controlled current modulation (such as sinewave modulation) may be necessary in the control at low speeds.

### 7.8 Solid rotors

In the conventional SR motor both rotor and stator are laminated; induced currents in either member would generally impair the torque production and produce additional losses. If there is no magnetic saturation the instantaneous torque is given by

$$T = \tfrac{1}{2} i^2 \frac{\mathrm{d}L}{\mathrm{d}\theta}.$$

Suppose that the rotor is made solid, or short-circuited 'shading coils' are affixed to its poles. If the rotor resistance is negligible the torque is given by the same expression but with $L$ replaced by

$$L' = L(1 - k^2)$$

where $k$ is the coupling coefficient between the stator winding and the rotor circuit. $L'$ is the leakage inductance and is completely defined by this equation. It is, of course, meaningful only when the stator flux is changing at a sufficiently rapid rate to ensure that the induced currents in the rotor are 'inductance limited'; this, however, is not difficult to achieve. Figure 7.25 shows an experimental rotor of this type.

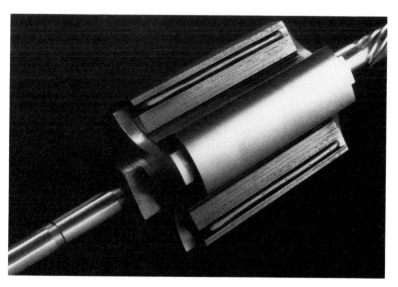

FIG. 7.25. Experimental solid-rotor switched repulsion motor.

With suitable pole geometry the minimum value of $L'$ occurs when the stator and rotor poles are aligned, and the maximum value is when they are unaligned. Interestingly, the current pulses must now be phased to coincide with the separation of the poles, and not with their approach. The torque is produce by repulsion, not by attraction, so this is essentially a switched repulsion motor.

## Problems for Chapter 7

1. What is the step angle of a three-phase switched reluctance motor having 12 stator poles and 8 rotor poles? What is the commutation frequency in each phase at a speed of 6000 r.p.m.?

2. What is the step angle of a five-phase switched reluctance motor having 10 stator poles and 4 rotor poles? What is the commutation frequency in each phase at a speed of 6000 r.p.m.?

3. A switched reluctance motor with six stator poles and four rotor poles has a stator pole arc of 30° and a rotor pole arc of 32°. The aligned inductance is 10.7 mH and the unaligned inductance is 1.5 mH. Saturation can be neglected. Calculate the instantaneous torque when the rotor is 30° before the aligned position and the phase current is 7 A. Neglect fringing.

4. In the motor of Problem 3, what is the maximum energy conversion in one stroke if the current is limited to 7.0 A? Determine the average torque corresponding to this energy conversion.

5. In the motor of Problem 3, what is the flux-linkage in the aligned position when phase current is 7.0 A? If this flux-linkage can be maintained constant while the rotor rotates from the unaligned position to the aligned position at low speed, determine the energy conversion per stroke and the average torque.

6. Show that for an unsaturated switched reluctance motor operating with a fixed conduction angle and flat-topped current waveform, the average torque is proportional to $V^2/\omega^2$ where $V$ is the supply voltage and $\omega$ is the angular velocity. Hence show that to maintain constant torque per ampere it is necessary to maintain the 'volts per Hertz' constant. Also show that with fixed supply voltage, a constant-power

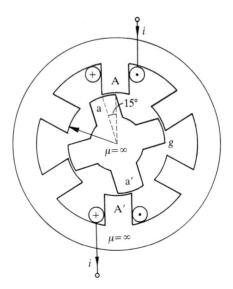

Fig. 7.26.

characteristic can be obtained by making the conduction angle proportional to the speed.

7. Figure 7.26 shows the cross-section of a switched reluctance motor. The rotor is in such a position that the 'overlap angle' between poles aa' of the rotor and poles AA' of the stator is 15°. The airgap $g = 0.5$ mm, stator bore radius $r = 25$ mm, and axial length $l = 50$ mm. There are 98 turns on each stator pole.
   (a) A current of 6 A flows through the two coils on poles A and A' in series. Sketch the flux paths on Fig. 7.26 for this condition, showing six flux-lines.
   (b) Calculate the flux-density in the airgap between the poles a, A for the conditions in part (a).
   (c) For the conditions in part (a), calculate the torque and indicate its direction. Neglect fringing and leakage, and assume that the steel parts are infinitely permeable.
   (d) Through what angle of rotation is the torque essentially constant, if the current is constant and there is no fringing?

# References and further reading

Acarnley, P. P. (1982). *Stepping motors: a guide to modern theory and practice*. Peter Peregrinus, London.

Acarnley, P. P., Hill, R. J., and Hopper, C. W. (1985). Detection of rotor position in stepping and switched motors by monitoring of current waveforms. *IEEE Transactions*, **IE-32**, pp. 215–22.

Baliga, B. J. (1987). *Modern power devices*. Wiley, New York.

Bartlett, P. M. (1984). The development of design specifications for brushless d.c. servomotors. *Drives/Motors/Controls*, pp. 92–100.

Bass, J. T., Miller, T. J. E., and Ehsani, M. (1986). Robust torque control of switched-reluctance motors without a shaft-position sensor. *IEEE Transactions*, **IE-33**, pp. 212–16.

Bass, J. T., Ehsani, M., and Miller, T. J. E. (1987a). Simplified electronics for torque control of sensorless switched reluctance motor. *IEEE Transactions*, **IE-34**, pp. 234–9.

Bass, J. T., Ehsani, M., Miller, T. J. E., and Steigerwald, R. (1987b). Development of a unipolar converter for variable-reluctance motor drives. *IEEE Transactions*, **IA-23**, 545–53.

Bausch, H. and Rieke, B. (1976). Speed and torque control of thyristor-fed reluctance motors. In *Proceedings of the International Conference on Electrical Machines, Vienna*, Part I, 128.1–128.10.

Bausch, H. and Rieke, B. (1978). Performance of thyristor-fed electric car reluctance machines. In *Proceedings of the International Conference on Electrical Machines, Brussels*, E4/2.1–2.10.

Bedford, B. D. (1972a). US Patent 3,678,352.

Bedford, B. D. (1972b). US Patent 3,679,953.

Binns, K. J. (1984). Permanent-magnet motors for inverter-fed drives. *Drives/Motors/Controls*, 101–5.

Binns, K. J. and Kurdali, A. (1979). Permanent-magnet a.c. generators. *Proceedings IEE*, **126**, 690–6.

Binns, K. J., Barnard, W. R., and Jabbar, M. A. (1978). Hybrid permanent-magnet synchronous motors. *Proceedings IEE*, **125**, 203–8.

Blake, R. J., Davis, R. M., Ray, W. F., Fulton, N. N., Lawrenson, P. J., and Stephenson, J. M. (1984). The control of switched reluctance motors for battery electric road vehicles. In *IEE Conference Publication No. 234, Power Electronics and Variable-Speed Drives*, pp. 361–4.

Blake, R. J., Webster, P. D., and Sugden, D. M. (1986). The application of GTOs to switched reluctance drives. In *IEE Conference Publication No. 264, Power Electronics and Variable-Speed Drives*, pp. 24–8. Institution of Electrical Engineers, London.

Blaschke, F. (1972). The principle of field orientation as applied to the new Transvector closed-loop control system for rotating field machines. *Siemens Review*, **34**, pp. 217–21.

Bolton, H. R. and Ashen, R. A. (1984). Influence of motor design and feed current waveforms on torque ripple in brushless DC drives. *Proceedings IEE*, Pt. B, **131**, (3), 82–90.

Bolton, H. R., Liu, Y. D., and Mallinson, N. M. (1986). Investigation into a class of brushless d.c. motors with quasi-square voltages and currents. *Proceedings IEE*, **133**, Pt. B, 103–11.

Bose, B. K. (1986). *Power electronics and AC drives*. Prentice-Hall, Englewood Cliffs, NJ, USA.

Bose, B. K. (1987). *Microcomputer control of AC drives*. IEEE Press, NY, USA.

Bose, B. K., Miller, T. J. E., Szczesny, P. M., and Bicknell, W. H. (1986). Microcomputer control of switched reluctance motor. *IEEE Transactions*, **IA-22**, 708–15.

Byrne, J. V. (1972). Tangential forces in overlapped pole geometries incorporating ideally saturable material. *IEEE Transactions*, **MAG-8**, pp. 2–9.

Byrne, J. V. and Lacy, J. G. (1976). Characteristics of saturable stepper and reluctance motors. In *IEE Conference Publication No. 136, Small Electrical Machines*, pp. 93–6.

Byrne, J. V. and McMullin, M. F. (1982). Design of a reluctance motor as a 10 kW spindle drive. *Motorcon Proceedings*, pp. 10–24.

Byrne, J. V. and O'Dwyer, J. B. (1976). Saturable variable-reluctance machine simulation using exponential functions. In *Proceedings of the International Conference on Stepping Motors and Systems, Leeds, 13–15 July, 1976*, pp. 11–16.

Byrne, J. V., O'Dwyer, J. B., and McMullin, M. F. (1985). A high-performance variable reluctance drive: a new brushless servo. *Motorcon Proceedings*, pp. 147–60.

Chalmers, B. J., Hamed, S. A., and Baines, G. D. (1985). Parameters and performance of a high-field permanent-magnet synchronous motor for variable-frequency operations. *Proceedings IEE*, **132**, Pt. B, pp. 117–24.

Chan, C. C. (1987). Single-phase switched reluctance motors. *Proceedings IEE*, **134**, Pt. B, pp. 53–6.

Chappell, P. H., Ray, W. F., and Blake, R. J. (1984). Microprocessor control of a variable reluctance motor. *Proceedings IEE*, **131**, Pt. B, pp. 51–60.

Chira, A. and Fukao, T. (1987). A closed loop control of super high-speed reluctance motor for quick torque response. In *IEEE Industry Applications Society Annual Meeting, Atlanta, October 1987*.

Colby, R. S. and Novotny, D. W. (1987). Efficient operation of surface-mounted PM synchronous machines. *IEEE Transactions*, **IA-23**, pp. 1048–54.

Corda, J. and Mateljan, T. (1987). Optimization of operation of a switched reluctance motor using nonlinear mathematical model. In *IEEE Conference Publication No. 282, Electrical Machines and Drives*, pp. 352–4.

Cornell, E. P. (1983). Permanent magnet AC motors. In *Drives/Motors/Controls 83 Conference Proceedings*, pp. 102–7. Peter Peregrinus, London.

Cruickshank, A. J. O., Anderson, A. F., and Menzies, R. W. (1966). Axially laminated anisotropic rotors for reluctance motors. *Proceedings IEE*, **113**, pp. 2058–60.

Davis, R. M. (1983). The switched reluctance drive. *Drives/Motors/Controls*, pp. 188–91.

Davis, R. M., Ray, W. F., and Blake, R. J. (1981). Inverter drive for switched reluctance motor; circuits and component ratings. *Proceedings IEE*, **128**, Pt. B, pp. 126–36.

Dawson, G. E., Eastham, A. R., and Mizia, J. (1987). Switched-reluctance motor torque characteristics: finite-element analysis and test results. *IEEE Transactions*, IA-23, pp. 532–7.

Demerdash, N. A., Miller, R. H., Nehl, T. W., Overton, B. P., and Ford, C. J. (1983). Comparison between features and performance characteristics of fifteen-hp samarium-cobalt and ferrite-based brushless d.c. motors operated by same power conditioner. *IEEE Transactions*, **PAS-102**, pp. 104–12.

Demerdash, N. A., Nehl, T. W., Fouad, F. A., and Arkadan, A. A. (1984). Analysis of the magnetic field in rotating-armature electronically commutated d.c. machines by finite elements. *IEEE Transactions*, **PAS-103**, pp. 1829–36.

Demerdash, N. A., Nehl, T. W., and Nyamusa, T. A. (1985). Comparison of effects of overload on parameters and performance of samarium-cobalt and strontium-ferrite PM brushless d.c. motors. *IEEE Transactions*, **PAS-104**, pp. 2223–31.

Dewan, S. B., Slemon, G. R., and Straughen, A. (1984). *Power semiconductor drives.* John Wiley—Interscience, New York.

Finch, J. W. and Lawrenson, P. J. (1978). Synchronous performance of single-phase reluctance motors. *Proceedings IEE*, **125**, pp. 1350–6.

Finch, J. W. and Lawrenson, P. J. (1979). Asynchronous performance of single-phase reluctance motors. *Proceedings IEE*, **126**, pp. 1249–54.

Finch, J. W., Harris, M. R., Musoke, A., and Metwally, H. M. B. (1984). Variable speed drives using multi-tooth per pole switched reluctance motors. In *13th Incremental Motion Control Systems Symposium, Urbana-Champaign, Illinois*, pp. 293–302.

Finch, J. W., Metwally, H. M. B., and Harris, M. R. (1986). Switched reluctance motor excitation current: scope for improvement. In *IEE Conference Publication No. 264, Power Electronics and Variable-Speed Drives*, pp. 196–9.

Fong, W. and Htsui, J. S. C. (1970). New type of reluctance motor. *Proceedings IEE*, **117**, pp. 545–51.

French, P. and Williams, A. H. (1967). A new electric propulsion motor. In *Proceedings AIAA Third Propulsion Joint Specialist Conference*.

Fukao, T. (1986). Principles and output characteristics of super high speed reluctance generator system. *IEEE Transactions*, **IA-22**, pp. 702–7.

Fulton, N. N. (1987). The application of CAD to switched reluctance drives. In *IEE Conference Publication No. 282, Electrical Machines and Drives*, pp. 275–9.

Geiger, D. F. (1981). *Phaselock loops for DC motor speed control.* John Wiley—Interscience, New York.

Harris, M. R. (1975). Static torque production in saturated doubly-salient machines. *Proceedings IEE*, **122**, pp. 1121–7.

Harris, M. R. (1982). Brushless motors—a selective review. *Drives/Motors/Controls*, pp. 84–90.

Harris, M. R. and Andjargholi, V. (1974). Limitations on reluctance torque in doubly-salient structures. In *Proceedings of the International Conference on Stepping Motors and Systems, Leeds, 15–18 July*, pp. 158–67.

Harris, M. R., Andjargholi, V., Lawrenson, P. J., Hughes, A., and Ertan, B. (1977). Unifying approach to the static torque of stepping motor structures. *Proceedings IEE*, **124**, pp. 1215–24.

Harris, M. R., Finch, J. W., Mallick, J. A., and Miller, T. J. E. (1986). A review of the integral-horsepower switched reluctance drive. *IEEE Transactions*, **IA-22**, pp. 716–21.

Himei, T., Funabiki, S., Agari, Y., and Okada, M. (1985). Analysis of voltage source inverter-fed permanent-magnet synchronous motor taking account of converter performance. *IEEE Transactions*, **IA-21**, pp. 279–84.

Honsinger, V. B. (1971a). The inductances Ld and Lq of reluctance machines. *IEEE Transactions*, **PAS-90**, pp. 298–304.

Honsinger, V. B. (1971b). Steady-state performance of reluctance machines. *IEEE Transactions*, **PAS-90**, pp. 305–17.

Honsinger, V. B. (1972). Inherently stable reluctance motors having improved performance. *IEEE Transactions*, **PAS-91**, pp. 1544–53.

Honsinger, V. B. (1982). The fields and parameters of interior type AC permanent magnet machines. *IEEE Transactions*, **PAS-101**, pp. 867–75.

Hughes, A. and Miller, T. J. E. (1977). Analysis of fields and inductances in air-cored and iron-cored synchronous machines. *Proceedings IEE*, **124**, pp. 121–8.

Jahns, T. M. (1984). Torque production in PM synchronous motor drives with rectangular current excitation. *IEEE Transactions*, **IA-20**, pp. 803–13.

Jahns, T. M. (1987). Flux-weakening regime operation of an interior permanent magnet synchronous motor drive. *IEEE Transactions*, **IA-23**, pp. 681–9.

Jahns, T. M., Kliman, G. B., and Neumann, T. W. (1986). Interior magnet synchronous motors for adjustable-speed drives. *IEEE Transactions*, **IA-22**, pp. 738–47.

Jarret, J. (1976). Machines electriques a reluctance variable et a dents saturees. *Tech. Mod.* **2**, pp. 78–80.

Jordan, H. (1983). *Energy efficient electric motors and their application*. Van Nostrand Reinhold, New York.

Kaufman, G., Garces, L., and Gallagher, G. (1982). High-performance servo drives for machine-tool applications. IEEE IAS Annual Meeting, pp. 604–9.

Kazdaghli, A., Razek, A., and Faure, E. (1983). Utilization des aimants permanents dans les machines synchrones a vitesse variable et elevee. (Use of permanent magnets in high-speed synchronous motor drives). *RGE*, **May 1983**, pp. 337–42.

Kenjo, T. and Nagamori, S. (1985). *Permanent-magnet and brushless d.c. motors*. Clarendon Press, Oxford.

Kenjo, T. (1985). *Stepping motors and their microprocessor controls*. Clarendon Press, Oxford.

Koch, W. H. (1977). Thyristor controlled pulsating field reluctance motor system. *Electric Machines and Electromechanics*, **1**, pp. 201–15.

Kuo, B. C. (1979). *Step motors and control systems*. SRL Publishing Company, Champaign, Illinois, USA.

Kusko, A. and Peeran, S. Y. (1987). Brushless DC motors using unsymmetrical field magnetization. *IEEE Transactions*, **IA-23**, pp. 319–26.

Lajoie-Mazenc, M. *et al.* (1976). An electrical machine with electronic commutation using high energy ferrite. In *Proceedings of IEE Small Electrical Machines Conference*, pp. 31–4.

Lajoie-Mazenc, M. *et al.* (1983). Feeding permanent-magnet machines by a transistorized inverter. *PCI/Motorcon*, pp. 558–70.

Lajoie-Mazenc, M., Mathieu, P., and Davat, B. (1984a). Utilization des aimants permanents dans les machines a commutation electronique. *RGE*, pp. 605–12.

Lajoie-Mazenc, M., Villanueva, C., and Hector, J. (1984b). Study and implementation of hysteresis controlled inverter on a permanent magnet synchronous machine. *Transactions IEEE*, **IA-21**, pp. 426–31.

Lang, J. H. and Vallese, F. J. (1985). *Variable reluctance motor drives for electric vehicle propulsion.* US D.o.E. Report DOE/CS-54209-26, 1 May, 1985.

Langley, L. W. and Lacey, R. J. (1981). Electric main propulsion drive for remotely piloted vehicle. *IEE Conference Publication No. 202, Small and Special Electrical Machines,* pp. 158–61.

Lawrenson, P. J. (1983). Switched reluctance motor drives. *Electronics and Power,* **February 1983,** pp. 144–7.

Lawrenson, P. J. and Agu, L. A. (1964). Theory and performance of polyphase reluctance machines. *Proceedings IEE,* **111,** pp. 1435–45.

Lawrenson, P. J. and Bowes, S. R. (1971). Stability of reluctance machines. *Proceedings IEE,* **118,** pp. 356–69.

Lawrenson, P. J. and Gupta, S. K. (1967). Developments in the theory and performance of segmental-rotor reluctance machines. *Proceedings IEE,* **114,** pp. 645–53.

Lawrenson, P. J. and Gupta, S. K. (1971). Fringe and permeance factors for segmented-rotor reluctance machines. *Proceedings IEE,* **118,** pp. 669–73.

Lawrenson, P. J. and Mathur, R. M. (1973). Pull-in criterion for reluctance motors. *Proceedings IEE,* **120,** 982–5.

Lawrenson, P. J., Mathur, R. M., and Vamaraju, S. R. M. (1969). Importance of winding and permeance harmonics in the prediction of reluctance-motor performance. *Proceedings IEE,* **116,** pp. 781–7.

Lawrenson, P. J., Mathur, R. M., and Stephenson, J. M. (1971). Transient performance of reluctance machines. *Proceedings IEE,* **118,** 777–83.

Lawrenson, P. J., Stephenson, J. M., Blenkinsop, P. T., Corda, J., and Fulton, N. N. (1980). Variable-speed switched reluctance motors. *Proceedings IEE,* **127,** Pt. B, pp. 253–65. (See also Discussion, *ibid.,* pp. 260–8.)

Lawrenson, P. J., Ray, W. F., Davis, R. M., Stephenson, J. M., Fulton, N. N., and Blake, R. J. (1982). Controlled-speed switched reluctance motors—present status and future potential. *Drive/Motors/Controls,* pp. 23–31.

Leonhard, W. (1985). *Control of electrical drives.* Springer-Verlag, Berlin.

Le-Huy, H., Perret, R., and Feuillet, R. (1986). Minimization of torque ripple in brushless d.c. motor drives. *IEEE Transactions,* **IA-23,** pp. 748–55.

Levi, E. (1984). *Polyphase Motors.* John Wiley & Sons, New York.

Lumsdaine, A., Lang, J. H., and Balas, M. J. (1986). A state observer for variable reluctance motors. In *Proceedings of the Incremental Motion Control Systems Symposium, June 3–5, 1986,* pp. 267–73, Urbana-Champaign, Illinois, USA.

Mhango, L. M. C. and Creighton, G. K. (1984). Novel two-phase inverter-fed induction-motor drive. *Proceedings IEE,* **131,** pp. 99–104.

Miller, T. J. E. (1984). Synchronization of line-start permanent-magnet AC motors, *IEEE Transactions,* **PAS-103,** pp. 1822–8.

Miller, T. J. E. (1985). Converter volt-ampere requirements of the switched reluctance drive, *IEEE Transactions,* **IA-21,** pp. 1136–44.

Miller, T. J. E. (1987a). Small motor drives expand their technology horizons. *IEE Power Engineering Journal,* **1,** pp. 283–9.

Miller, T. J. E. (1987b). Brushless reluctance motor drives. *IEEE Power Engineering Journal,* **1,** pp. 325–31.

Miller, T. J. E. (1988). Brushless permanent-magnet motor drives. *IEE Power Engineering Journal,* **2,** pp. 55–60.

Miller, T. J. E. and Hughes, A. (1977). Comparative design and performance analysis of air-cored and iron-cored synchronous machines. *Proceedings IEE*, **124**, pp. 127–32.

Miller, T. J. E. and Jahns, T. M. (1986). A current-controlled switched reluctance drive for FHP applications. In *Conference on Applied Motion Control, Minneapolis, USA, June 10–12*, pp. 109–17.

Miller, T. J. E. and McGilp, M. (1987). PC CAD for switched reluctance drives. In *IEE Conference on Electrical Machines and Drives, Conference Publication No. 282, London, November 16–18*, pp. 360–6.

Murphy, J. M. D. (1973). *Thyristor control of AC motors.* Pergamon Press, Oxford.

Nasar, S. A. (1969). DC switched reluctance motor. *Proceedings IEE*, **116**, (6), 1048–9.

Nehl, T., Fouad, F., and Demerdash, N. (1982). Dynamic simulation of radially oriented permanent magnet type electronically operated synchronous machines with parameters obtained from finite-element field solutions. *IEEE Transactions*, **IA-18**, pp. 172–82.

Nehl, T. *et al.* (1985a). Automatic formulation of models for simulation of the dynamic response of electronically commutated d.c. motors. *IEEE Transactions*, **PAS-104**, pp. 2214–22.

Nehl, T., Demerdash, N., Fouad, F. (1985b). Impact of winding inductances and other parameters on the design and performance of brushless d.c. motors. *IEEE Transactions*, **PAS-104**, pp. 2206–13.

Owyang, K., Yilmaz, H., and Wildi, E. (1987). High-voltage switching module uses power IC technology. *Powertechnics*, **Jan. 87**, pp. 25–30.

Patni, C. K. and Williams, B. W. (1986). The effect of modulation techniques and electromagnetic design on torque ripple in brushless PM motors. *IEE Conference Publication No. 264, Power Electronics and Variable-Speed Drives*, pp. 76–9.

Pollock, C. and Williams, B. W. (1987). *An integrated approach to switched reluctance motor design.* European Power Electronics Conference, Grenoble.

Plunkett, A. B. (1979). *Current controlled PWM transistor inverter drive.* IEEE IAS Annual Meeting, pp. 785–92.

Ray, W. F. and Davis, R. M. (1979). Inverter drive for doubly salient reluctance motor: its fundamental behaviour, linear analysis, and cost implications. *Electric Power Applications*, **2**, pp. 185–93.

Ray, W. F. and Davis, R. M. (1981). Inverter drive for switched reluctance motor: circuits and component ratings. *Proceedings IEE*, **128**, Pt. B, pp. 126–36.

Ray, W. F., Davis, R. M., Stephenson, J. M., Lawrenson, P. J., Blake, R. J., and Fulton, N. N. (1984). Industrial switched reluctance drives—concepts and performance. *Drives/Motors/Controls*, pp. 357–60.

Ray, W. F., Lawrenson, P. J., Davis, R. M., Stephenson, J. M., Fulton, N. N., and Blake, R. J. (1986a). High-performance switched reluctance brushless drives. *IEEE Transactions*, **IA-22**, pp. 722–30.

Ray, W. F., Davis, R. M., and Blake, R. J. (1986b). The control of SR Motors. In *Conference on Applied Motion Control, Minneapolis, USA, June 10–12*, pp. 137–45.

Regas, K. A. and Kendig, S. D. (1987). Step motors that perform like servos. *Machine Design*, **December 10**, pp. 116–20.

Richardson, K. M. and Spooner, E. (1987). Magnetization procedures for Nd-Fe-B magnets in electrical machines. In *IEE Conference Publication No. 282, Electrical Machines and Drives*, pp. 250–64.

Richter, E. and Neumann, T. W. (1981). Synchronous machine designs using different types of permanent magnets. In *Fifth International Workshop on Rare Earth Cobalt Magnets and their Applications, Roanoke, Virginia, June 1981.*

Richter, E. and Neumann, T. W. (1984). Saturation effects in salient pole synchronous motors with permanent magnet excitation. In *International Conference on Electrical Machines, Lausanne, Switzerland, September 1984.*

Richter, E., Miller, T. J. E., Neumann, T. W., and Hudson, T. L. (1985). The ferrite permanent-magnet a.c. motor: a technical and economic assessment. *IEEE Transactions,* **IA-21,** pp. 644–50.

Ross, J. S. H. (1971). UK Patent 1 395 152.

Sebastian, T. and Slemon, G. R. (1987). Operating limits of inverter-driven permanent magnet motor drives, *IEEE Transactions,* **IA-23,** pp. 327–33.

Semail, B., Piriou, F. and Razek, A. (1987). Numerical model for PM synchronous motor drives to limit torque pulsations. In *IEE Conference Publication No. 282, Electrical Machines and Drives,* pp. 348–51.

Sen, P. C. (1981). *Thyristor DC drives.* John Wiley,Interscience, New York.

Shanliang, You *et al.* (1983). Large-capacity generators using rare-earth cobalt permanent magnets. In *Proceedings of the Seventh International Workshop on Rare-Earth Cobalt Permanent Magnets and their Applications,* pp. 21–8.

Sitzia, A. M. and Chalmers, B. J. (1987). Electromagnetic design of brushless d.c. motor with slotless stator. *IEE Conference Publication No. 282, Electrical Machines and Drives,* pp. 260–4.

Slemon, G. R. and Gumaste, A. V. (1983). Steady-state analysis of a permanent-magnet synchronous motor drive with current-source inverter. *IEEE Transactions,* **IA-19,** pp. 190–7.

Sneyers, B., Novotny, D. W., and Lipo, T. A. (1985). Field weakening in buried permanent-magnet AC motor drives. *IEEE Transactions,* **IA-21,** pp. 398–407.

Steigerwald, R. L., Kuo, M. H., and Claydon, G. S. (1987). HVIC for power supply applications. *Powertechnics,* **June 1987,** pp. 21–5.

Stephenson, J. M. and Corda, J. (1979). Computation of torque and current in doubly salient reluctance motors from nonlinear magnetization data. *Proceedings IEE,* **126,** pp. 393–6.

Stephenson, J. M. and Lawrenson, P. J. (1969). Average asynchronous torque of synchronous machines, with particular reference to reluctance machines. *Proceedings IEE,* **116,** pp. 1049–51.

Stone, A. C. and Buckley, M. G. (1984). Ultra high performance brushless d.c. drive. *Drives/Motors/Controls,* pp. 86–91.

Torrey, D. A. and Lang, J. H. (1986). A GTO based inverter for a 60 kW variable reluctance motor drive for electric vehicle propulsion. In *International Electric Vehicle Symposium, Washington D.C., October 20–23.*

Unnewehr, L. E. and Koch, W. H. (1974). An axial air-gap reluctance motor for variable-speed applications. *IEEE Transactions,* **PAS-93,** pp. 367–76.

Viarouge, P., Lajoie-Mazenc, M., and Andrieux, C. (1987). Design and construction of a brushless permanent-magnet servomotor for direct-drive application. *IEEE Transactions,* **IA-23,** pp. 526–31.

Volkrodt, W. (1976). Machines of medium-high rating with a ferrite-magnet field. *Siemens Review,* **43,** pp. 248–54.

Webster, P. D. (1986). Isolated current measurement for the switched reluctance drive.

In *IEE Conference Publication No. 264, Power Electronics and Variable-Speed Drives*, pp. 177–81.

Weinmann, D., Nicoud, G., and Gallo, F. (1984). Advantages of permanent magnet motors. *Drives/Motors/Controls*, pp. 113–20.

Wiesemann, R. (1927). Graphical determination of magnetic fields. *AIEE Transactions*, **46,** pp. 141–54.

Wildi, E. J. *et al.* (1984). New high-voltage IC technology. *IEEE International Electron Devices Meeting, paper 10.2*, pp. 262–5.

# Answers to problems

## Chapter 2

1. 36.4 mm.
2. 14.4 kN m/m³; 1.04 p.s.i.
3. 3.6 A/mm².
4. $4.49 \times 10^{-3}$ kg m²; 3.35; 5223 rad/s².
5. 8.33 N-m.

## Chapter 3

2. (b).
3. (b).
4. (a).
5. (a) stator; (b) rotor.
6. $\int H . dl = Ni$.
7. (a), (c).
9. $B_m H_m$.
10. 398 J.
11. (c).
12. 50 per cent.
13. 1.14 T.
14. 0.9 T.
15. $-80$ kA/m or $-1$ kOe.
16. 0.278 T; 199 kJ/m³ (25 MGOe); 35 kJ/m³ (4.4 MGOe).

## Chapter 4

1. Yes.
2. (c) $7.45 \times 10^{-1}$ Wb; (d) 0.275 T; (e) $-174$ At, $-21.7$ kA/m; (f) 22.3 mV s; (h) 7.26 V; (i) 0.36 mH; (j) 953 A, not practical; (k) 0.223 J.
3. (a) 3820 r.p.m.; (b) 153 A, 18.4 N m; (c) 84 per cent; (i) 16.4 W, (ii) 20.2 W, (iii) 16.4 W, (iv) 10.6 W; 99.8 Ω.
4. 2836 r.p.m.
5. 0.225 T; $1.13 \times 10^{-3}$ N m.
6. 1337 r.p.m.; 1337 r.p.m.

## Chapter 5

1. 173 V.
2. 9.73 Ω.

# ANSWERS TO THE PROBLEMS

3. 6.6 N m.
4. 195 V; 15.3°.
5. 206 V; 27.9°.
6. 8260 r.p.m.

## Chapter 6

1. 0.921 N m.
2. 20.29°; 0.313 N m.
3. 3668 r.p.m.
4. $V=42.75$ V; $\delta=28.3°$; $T=1.89$ N m; PF$=0.73$ lag; $P=595$ W; $VA/W=1.15$; $\sigma=2.07$ p.s.i.; $T/I=0.237$ N m/A.
5. 1.1 Ω, 2.9 Ω; 0.093 N m.

## Chapter 7

1. 15°; 800 Hz.
2. 18°; 400 Hz.
3. 0.43 N m.
4. 0.225 J; 0.43 N m.
5. 74.9 mV s; 1.6 J; 3.07 N m.
7. 1.48 T; 1.09 N m; 30°.

# Index

acceleration 25 ff.
　controlled 3
ACSL 10, 27
a.c. commutator motor 11, 156
a.c. drives 4, 8, 16
actuator 3, 6, 33
aerospace 6
air conditioning 5, 6
airgap length 18, 55, 112
airgap flux 54, 85; *see also* flux-density, airgap
airgap shear stress 20 ff.
　typical values 24
alignment torque 29
Alnico 34, 37, 41
Ampere's law 39, 94
Anderson Strathclyde plc 5
angular velocity 92
armature reaction 13
automotive 6, 13
axis drives 2

Bedford, B.D. 149
B-H loop 35
braking 4
brushes 54–5
　maintenance 54
brushgear 13
brushless d.c. motor 1, 12, 14, 17, 32, 173, 54 ff.
　evolution of 12 ff.
brushless d.c. motor, sinewave 17, 88
　airgap flux-density 88, 92
　ampere-conductor distribution, rotating 91, 102, 116
　armature reaction 91
　　flux per pole 95
　back-e.m.f. 102, 109
　circle diagram 103 ff.
　computer simulation 112
　conductor distribution 89
　configuration 88 ff.
　control 103 ff.
　current regulation 2
　current waveform 110
　efficiency 112, 114
　e.m.f. equation 92, 99
　flux distribution 88, 91
　flux, rotating 93, 96
　flux per pole 93, 95, 99, 110
　　magnet 93–4
　　armature reaction 95
　harmonics 99, 116–17
　inductance of phase winding 94
　kVA requirement 103, 109 ff.
　leakage reactance 100
　magnet pole arc 88
　m.m.f. distribution 94
　operation 138 ff.
　phase winding 94 ff.
　　inductance 94
　phasor diagram 100 ff.
　pole arc 88
　power factor 103, 107
　sine-distributed turns 90, 97
　slots per pole per phase 95, 98
　speed range 107
　star connection 109
　synchronous reactance 100
　　of ideal sinewave motor 96
　torque angle 91, 99–100, 102, 109
　torque dip 74
　torque equation 89 ff., 99, 104
　torque/speed characteristic 103 ff.
　torque per ampere 109 ff.
　torque ripple 89, 96, 116–17
　windings 88, 91, 96
　winding factors 96 ff.
brushless d.c. motor, squarewave 32
　airgap flux-density 61–2, 86
　armature constant 66
　armature reaction 76
　back-e.m.f. 74
　CAD 85
　commutation 66, 70 ff., 82
　commutation tables 72
　computer simulation 83 ff.
　configuration 54–5
　control 80 ff.
　current regulation 82
　current waveform 73–4
　efficiency 55, 66, 86
　e.m.f. equation 63 ff., 68–70
　inductance 76 ff.
　kVA requirement 109 ff.
　magnet operating point 61
　magnetic circuit analysis 58 ff., 85
　magnetization of magnets 86
　m.m.f. distribution 74
　phasebelts 74

# INDEX

pole arc 63, 70 ff.
shaft position sensor 82
similarity with commutator motor 65–6, 82
slots per pole per phase 63, 77
speed range 68
squarewave 54 ff.
star and delta windings 70–6
torque dip 74
torque equation 63, 65, 68–70
torque ripple 76, 79–80
torque/speed characteristic 66–8
120 and 180-deg. magnets 70 ff.
windings 63, 72
brushless exciter 17
brushless reluctance motor 149
Byrne, J.V. 149

CAD 9, 85
Carter's coefficient 60
ceramic magnets, see ferrite magnets
CGS units 41
chopping, see p.w.m.
circle diagram 103 ff., 145 ff.
classical motors 11 ff.
coal shearer 5
cobalt-samarium 34
coercivity, see permanent magnets
commutation 13, 32, 66
commutation tables 72
commutator 55
constant-power characteristic 119
control-C 10, 27
controlled acceleration 3
Control Techniques plc 5
'Crumax' 34
current density 22, 114, 138
current regulation 8, 82–3, 180 ff.
current sensors 10, 83

Darlington transistor 4
David McClure Ltd 5
D.C. commutator motor 5, 11 ff., 23, 32, 54, 87
digital control 8
digital electronics 7–8
direct-drive motor 149
disk drives 2, 3, 6
disk motor 148
distribution factor 98
doubly-excited machines 31, 149
doubly-salient machines 30
drive 1, 5
duty-cycle 85
dynamic braking 4

ECM (Electronically Commutated Motor) 56
efficiency 4, 16, 55, 66, 112, 118, 147, 152, 155
electric loading 21, 55
electronic commutation, see commutation
elevators 2
EMI 7
energy product, see permanent magnets
energy saving 1–2
environmental factors 7
EPROM 82
excitation penalty 19, 112
explosion proof enclosure 5

Faraday 93
Faults
  in PM motor 112
fault-tolerance 17
feasible triangle 156
ferrite magnets 9, 34, 35, 42, 46, 52
field control 13
field-oriented control 8, 16
field weakening 57, 112, 118–19
filter, RFI 54
finite-element analysis 9, 85, 167
floppy disk drives 2
flux concentration factor 119; see also magnetic circuit
flux-density
  airgap 46, 61–2, 90–1
  in permanent magnets 35, 61
flux concentration 119
flux focusing 119
flux-linkage, of sinewave motor 93
food processor 2
French, P. 149
frequency 92, 138, 156

gate array 6, 7, 82
Gauss' law 40
gearing 24 ff.
gear ratio 24 ff.
General Electric 9, 56, 149
GTO's 4

Hall-effect sensor 10–11, 82, 182
hand tools 2
hard disk drives 2
harmonics 3, 19, 54, 99, 116–17, 153
Harris, M.R. 149
Hewlett-Packard 149
high-efficiency motors 148
hybrid PM/synchronous reluctance motor 82; see also PM/synchronous reluctance hybrid motor

'hysteresisgraph' 37
hysteresis loop 35, 37
hysteresis motor 15

IGT/IGBT 5
induction motor 3, 11, 12, 14, 16, 19, 21, 22, 55, 82, 112, 150, 155
inductor motors 15
inertia loads 25 ff.
Inland-Kollmorgen 149
integrated circuits 83
interior magnet motor, *see* PM/synchronous reluctance hybrid motor

Jahns, T.M. 119, 143, 147
Jarret Company 149

keeper 36, 37, 39
Kevlar 9, 45, 55
Kliman, G.B. 119
Koch, W.H. 149
kVA requirement 103, 109 ff., 172

Lang, J.H. 149
Lawrenson, Prof. P.J. 120, 149 *et seq.*
leakage reactance 100
length/diameter ratio 24
Leonhard, Prof. W. 108
lifts 2
line-start motor 17, 135, 147
load angle 119; *see also* torque angle
load commutated converter 4
loading, electric 21, 55
loading, magnetic 21, 46
load sharing 5
loctite 45
loudspeaker 53
Lucas Engineering & Systems Ltd. 3, 167

machine tools 6, 8
Magnequench 34
magnetic circuit 37, 39, 58 ff., 85
  flux concentration factor 40
magnetic field intensity 35
magnetic loading 21, 46
magnetic poles 40
magnetic resonance (MR) 51
magnetic tape drives 2
Magnetics Research International 121
magnetization (J) 38
magnetizing fixtures 35
magnetizing force 35, 39
mechatronics 1

microprocessors 8
microprocessor control 180
MOSFETs 4, 9
motion control systems 1, 5
  structure of 5–6
motors
  applications 14–15
  evolution of 11 ff.
Multiplex wiring 13

Nasar, S.A. 149
naturally commutated converter 4
NdFeB, *see* neodymium-iron-boron
'NeIGT' 34
neodymium-iron-boron 9, 34, 37, 41
'Neomax' 34
Neumann, T.W. 119
noise 6, 7, 13, 153, 155

office machinery 6, 7
optical encoder 181
optical interrupter 11
Oulton drive 149
overlap

Pacific Scientific 56–7
paper mills 7
PC-SRD design software 185
peak torque 25–7
permanent magnets 31
  application of 46–7, 57–8
  assembly of large magnets 50
  coercive force 38
  coercivity 34
  cost 46
  Curie temperature 41
  demagnetization of 41
  demagnetization curve 37, 48–9, 51, 58
    knee of 38
  domain relaxation 43
  energy product 34, 38, 40
    maximum 39
    highest useable 47
  fixturing 45
  handling 44
  hard 36, 37
  irreversible losses 43–4
  long-term stability 43
  and large machines 112
  magnetization 38, 44, 46, 86
  mechanical properties 35, 44–6
  metallurgical change 42
  minimum volume of 40
  operating point 36, 39, 61, 134
    influence on temperature effects 43

# INDEX

production methods 34
properties 35
recoil permeability 37
remanence 34, 36
retention methods 45
reversible losses 41–2
safety 44–5
saturation 36
second-quadrant operation 37
shapes 45
soft 37
stored energy 38
temperature effects 31 ff., 47, 51
thermal expansion 35, 45
versus electromagnetic excitation 112 ff.
permeability, relative 39
permeance coefficient 37, 40, 61–2
permeance, rotor leakage 60–1
phasor diagram 100 ff.
PM motors 13, 150
  faults 112
  performance 155
  volume of magnet required 113
  vs. electrically-excited motors 112 ff.
  slotless 115–16
PM/synchronous reluctance hybrid motor 18, 89, 118 ff.
  airgap flux-density 119
  ampere-conductor distribution, 122, 128, 129, 131
  armature reaction 128
    flux per pole 129–30
  cage winding 118, 121–2, 147
  circle diagram 145 ff.
  conductor distribution 124, 125
  configuration 118
  constant-power characteristic 119, 143
  control 138 ff.
  converter volt-ampere requirement 144–5
  demagnetizing current 134–5
  efficiency 118, 147
  e.m.f. equation 127
  field weakening 118–19
  flux concentration 119
  flux distribution 126
  flux, rotating 119
  flux per pole 124, 126 ff.
    armature reaction 128, 129–30
    magnet 119
    open-circuit 127
  inductance of phase winding 122 ff.
  interior magnet configuration 118
  kVA requirement 143, 144
  leakage reactance 130–31
  line-start motor 135, 147
  laminations 121
  magnet configuration 126
  magnet operating point 134
  magnetic circuit analysis 122 ff.
  open-circuit e.m.f. 122 ff.
  operation 138–47
  parameters 132
  phasor diagram 135 ff.
  pole arc 132
  power factor 118, 139 ff.
  pure reluctance motor 137
  reluctance torque 119, 126
  short-circuit current 134
  sine-distributed turns 122
  speed range 118, 141
  synchronous reactance 121, 126, 128 ff., 137
  torque equation 136
  torque/speed characteristic 145 ff.
  torque per ampere 140, 143, 147
  two-phase motor 134, 137
  windings 122 ff.
  winding factors 127, 132
pitch factor 98
pole arc 63
position control 2, 6
power density 46
power factor 16
power integrated circuits (PICs) 8
printheads 2, 6
pump drives 6
p.w.m. (pulse-width modulation) 4–5, 13, 67, 82, 85
p.w.m. inverter 118

rare earth/cobalt magnets 41, 46, 58
refrigeration 6
relative permeability, see permeability, relative
reluctance motors 15
reluctance torque 29
remagnetization 41
remanence, see permanent magnets
repulsion motor, switched 188–9
resolver 181
resolver/digital converter 11
RFI (radio-frequency interference) 13, 54
ripple, see torque ripple
robot 2, 6, 149
rotating ampere-conductor distribution 91, 102
rotating flux 91–2, 96
rotating rectifier 17
rotors 118 ff.

safety, with magnets 44–5
saturation 91, 160, 164 ff., 168 ff.,
  of stator teeth 46
scaling effects 16, 57, 112

# INDEX

Scottish Power Electronics and Electric Drives (SPEED) 154
SCRs 3, 4, 57
Sensefet 9, 182
series control (of induction motor) 3
series a.c. commutator motor 11, 156
series resistance control 13
servo motor 1, 6, 32, 56, 188
shaft position sensing 10, 16, 18, 55, 82
shear stress, airgap 20 ff.
shoot-through 83, 150
SIMNON 9, 27
simulation 27
  of brushless d.c. motor 83 ff.
sine-distributed windings 88 f.
sinewave motor, *see* brushless d.c. sinewave motor; a.c. drives; PM and synchronous reluctance hybrid motors
singly-excited machines 29
six-step inverter 16
sizing 20
skew 116, 158
skew factor 98
slewing 188
slip 16
slotless motor 115–16
smartpower 83
soft starter 4
specific output 22
speed control 6
speed range 4, 17, 18
speed/torque characteristic, *see* torque/speed characteristic
spindle drive 6
split ratio 23
squarewave motor, *see* brushless d.c. squarewave motor
standards 7
stator 63
steel rolling mills 7
step angle 156
stepper motors 13, 15, 18, 31, 150, 153, 158
stress, shear 20 ff.
stroke 156
Sumitomo 9, 35
surface-magnet motor 10, 108, 138 ff., 143–4; *see also* brushless d.c. motors
switched reluctance motor drive 2, 4, 18, 82, 149 ff.
  airgap 153
  aligned position 157, 160
  applications 155
  bifilar winding 180
  cabling 155
  CAD 185
  commutation 163, 180 ff.
  constant-power characteristic 151–2
  control 180 ff.
  converter circuits 150, 173 ff.
  converter volt-ampere requirement 172
  core losses 153
  current feedback 182
  current regulation 180 ff.
  current waveform 161–2, 166, 173
  design 185
  disadvantages 153
  efficiency 152, 155
  energy conversion loop 164, 172
  energy ratio 171
  fault conditions 150
  finite-element analysis 167
  frequency 156
  history 149
  inductance 160 ff.
  inductance ratio 157, 171
  kVA requirement 172
  low-speed operation 153, 173
  microprocessor control 180
  noise 153
  no. of motors per inverter 155
  $n+1$ switch circuit 177 *et seq.*
  open-circuit voltage 150
  operation 185
  overlap 156, 162
  PC-SRD design software 185 ff.
  phase number 156 ff.
  pole arc 156
  pole number 156 ff.
  pole shape 153, 158
  power density 155
  p.w.m. 178, 182
  ripple current, d.c. 29, 153
  rotor 150
  saturation 168 ff., 172
  shaft position sensor 155, 181, 188
  shoot-through immunity 150
  short-circuit current 150
  skew 158
  solid rotor 188
  speed range 150
  starting torque 150
  stator 150
  step angle 156
  steps/rev 156
  stroke 156
  torque 150, 159 ff.
  torque multiplication 153
  torque per ampere 152
  torque ripple 153, 155, 164
  torque per unit volume 153, 166
  torque/speed characteristic 150, 183 ff., 190
  unaligned position 156, 160
  unipolar converter circuits 173
  vernier principle 153, 172
  windings 150, 158
  working stroke, *see* stroke
Switched Reluctance Drives Ltd. 2, 149, 152

# INDEX

switched repulsion motor  188–9
synchronizing capability  148
synchronous machine  31, 121
synchronous motor  17, 88, 112, 188
synchronous reluctance motor  5, 17, 118 ff., 137 ff., 153; *see also* PM/synchronous reluctance hybrid motor
Synektron Corporation  3

TASC Drives Ltd.,  4, 149, 180
temperature effects, *see* permanent magnets
thyristor controlled drive  5
torque
 alignment  29, 118
 equation of brushless d.c. motor  63, 65
 reluctance  29, 118
torque/inertia ratio  4, 6, 32, 188
torque/speed characteristic  66 ff., 107, 183–4
torque per unit volume  20 ff., 166
torque production, basic principles  28 ff.
torque ripple  4, 11, 18, 19, 29 ff., 66, 116, 118, 155, 164
traction  5, 6, 112
transients  1
triac  3, 56
TRV  20 ff.
 typical values  24

two-phase motor  134, 137

under-excited operation  118–19
units  41
 conversion factors  41
universal motor, *see* series a.c. commutator motor
Unnewehr, L.E.  149

variable reluctance motor  149
vector control  8, 16
velocity control  2
voice-coil actuator  87
voltage-source inverter  147

Walter Jones & Co.  57
Warner Electric  149
washing machines  2
water cooling  5
waveforms, phase current  73
Williams, A.H.  149
windings  63, 72
winding factors, harmonic  96 ff.
wire-EDM  122
working stroke, *see* stroke

0.228 N m/A, and the terminal voltage is nearly 38 V. This is taken as the 'base' case for comparisons.

*Row 2:* What is the performance of the PM motor if the current is oriented in the q-axis, as it would normally be for the surface-magnet motor?

With the same current oriented along the q-axis, at the same speed, the torque decreases slightly to 0.912 N m. However, there is a small increase in the required voltage to 39.3 V, indicating that operation at this torque from a 38 V supply requires the field-weakening that results from advancing the current relative to the q-axis.

*Row 3:* What is the performance of the pure reluctance motor at maximum current and 3000 r.p.m.?

The reluctance motor has all the same parameters as the PM motor but $E_q = 0$. The optimum orientation for the current is at $\gamma = 45°$, giving a torque of only 0.066 N m and a specific torque of only 0.016 N m/A with maximum current flowing. However, the voltage required is only 8.75 V, implying that the reluctance motor could sustain this torque up to a higher speed. Note the very low power factor, 0.55 lagging, and the high volt-ampere requirement.

*Row 4:* Up to what speed could the reluctance motor sustain its maximum torque per ampere?

By solving the phasor equations with $I = 4.0$ A, $V = 37.6$ V, and $\gamma = 45°$, assuming that all the reactances are proportional to frequency, the frequency is found to be approximately 475 Hz and the speed 14 250 r.p.m. The power factor becomes even lower. Note that the 'volts per Hz' remained roughly constant in this calculation.

Evidently the reluctance torque is only about 7 per cent of the total torque of the PM motor, with the same voltage and current, but it must be borne in mind that these are small motors in which the resistance and leakage reactance are comparatively large. In a large motor both of these impedances would be smaller in relation to the synchronous reactances. An estimate of the expected relative improvement in larger motors can be made by assigning the resistance and leakage reactance to zero. While this is artificial, it permits all the other parameters to be kept the same, and provides a kind of 'per-unit' comparison.

*Row 5:* Reluctance motor with $R = X_\sigma = 0$

With 4.0 A oriented at 45° ahead of the q-axis, the torque remains the same because it depends on the difference between $X_d$ and $X_q$; the leakage reactance is common to both. However, the terminal voltage is now reduced to 5.36 V. The power factor actually decreases to 0.481 lagging, because of the removal of the in-phase voltage drop across the phase resistance. The efficiency of this hypothetical motor is 100 per cent, and even with no losses the volt-amperes per watt are still more than twice the value needed by the PM motor. At constant volts/Hz the torque could be maintained constant up to a speed of $38 \div 5.36 \times 3000 = 21\ 270$ r.p.m. without exceeding the current limit.

*Row 5a:* Reluctance motor with increased slot area and turns

The low voltage required by the reluctance motor is so far below the available

**Table 6.2.** Operation of PM and synchronous reluctance motors

| * | Motor | Parameter | $B_r$ | r.p.m. | $E_q$ | $I$ | $\gamma$ | $I_d$ | $I_q$ | $V_d$ | $V_q$ | $V$ | $\delta$ | $T$ | PF | $P$ | VA/W | $\sigma$ | $T/I$ |
|---|---|---|---|---|---|---|---|---|---|---|---|---|---|---|---|---|---|---|---|
| Row | | Unit | T | | V | A | ° | A | A | V | V | V | degrees | N m | | W | | p.s.i. | N m/A |
| 1 | PMH | | 1.1 | 3000 | 35.8 | 4.0 | 15 | −1.04 | 3.86 | −10.12 | 36.73 | 38.10 | 15.40 | 0.913 | 1.000 lg | 287.0 | 1.062 | 0.995 | 0.228 |
| 2 | PMH | | 1.1 | 3000 | 35.8 | 4.0 | 0 | 0.0 | 4.0 | −9.88 | 38.04 | 39.30 | 14.56 | 0.912 | 0.968 lg | 286.4 | 1.098 | 0.994 | 0.228 |
| 3 | REL | | — | 3000 | — | 4.0 | 45 | −2.83 | 2.83 | −8.58 | −1.76 | 8.75 | 101.6 | 0.066 | 0.551 lg | 20.7 | 3.38 | 0.072 | 0.016 |
| 4 | REL | | — | 14 250 | — | 4.0 | 45 | −2.83 | 2.83 | −34.79 | −14.28 | 37.60 | 112.3 | 0.066 | 0.386 lg | 98.1 | 3.07 | 0.072 | 0.016 |
| 5 | REL with $R=0$ and $X_\sigma=0$ | | — | 3000 | — | 4.0 | 45 | −2.83 | 2.83 | −5.15 | −1.50 | 5.36 | 106.2 | 0.066 | 0.481 lg | 20.7 | 2.61 | 0.072 | 0.016 |
| 5a | REL with increased slot area and turns | | — | 3000 | — | 4.0 | 45 | −2.83 | 2.83 | −35.95 | −11.77 | 37.83 | 108.1 | 0.304 | 0.452 lg | 95.5 | 3.17 | 0.332 | 0.076 |
| 6a | PMH | | 0.41 | 3000 | 11.3 | 4.0 | 0 | 0.0 | 4.0 | −9.88 | 13.58 | 16.79 | 36.04 | 0.289 | 0.807 lg | 90.7 | 1.48 | 0.315 | 0.072 |
| 6b | PMH | | 0.41 | 3000 | 11.3 | 4.0 | 15 | −1.04 | 3.86 | −10.12 | 12.28 | 15.92 | 39.50 | 0.312 | 0.910 lg | 98.0 | 1.30 | 0.340 | 0.078 |
| 6c | PMH | | 0.41 | 3000 | 11.3 | 4.0 | 25 | −1.69 | 3.63 | −9.90 | 11.38 | 15.08 | 41.03 | 0.312 | 0.961 lg | 98.0 | 1.23 | 0.340 | 0.078 |
| 6d | PMH | | 0.41 | 3000 | 11.3 | 4.0 | 45 | −2.83 | 2.83 | −8.57 | 9.59 | 12.86 | 41.79 | 0.270 | 0.998 ld | 84.8 | 1.21 | 0.294 | 0.068 |
| 7 | SPM | | 1.1 | 3000 | 47.6 | 4.0 | 0 | 0.0 | 4.0 | −4.09 | 49.84 | 50.01 | 4.69 | 1.212 | 0.997 lg | 381.0 | 1.05 | 1.322 | 0.303 |
| 8 | SPM | | 1.1 | 2250 | 35.7 | 4.0 | 0 | 0.0 | 4.0 | −3.07 | 37.94 | 38.06 | 4.62 | 1.212 | 0.997 lg | 285.6 | 1.07 | 1.322 | 0.303 |
| 9 | SPM | | 1.1 | 3000 | 47.6 | 38.1 | 70 | −35.8 | 13.0 | −33.3 | 18.29 | 38.02 | 61.24 | 3.940 | 0.988 ld | 1238 | 2.34 | 4.300 | 0.103 |
| 10 | SPM | | 0.41 | 3000 | 17.9 | 4.0 | 0 | 0.0 | 4.0 | −4.09 | 20.14 | 20.55 | 11.47 | 0.456 | 0.980 lg | 143.2 | 1.148 | 0.497 | 0.114 |
| 11 | SPM | | 0.41 | 5850 | 34.9 | 4.0 | 0 | 0.0 | 4.0 | −7.97 | 37.15 | 37.99 | 12.11 | 0.456 | 0.978 lg | 279.2 | 1.089 | 0.497 | 0.114 |
| 12 | SPM | | 0.41 | 7500 | 44.8 | 4.0 | 55 | −3.28 | 2.29 | −7.70 | 37.66 | 38.44 | 11.55 | 0.261 | 0.726 ld | 205.3 | 1.498 | 0.285 | 0.065 |
| 13 | PMH | | 0.41 | 8400 | 31.8 | 4.0 | 25 | −1.69 | 3.63 | −26.02 | 28.20 | 38.37 | 42.70 | 0.312 | 0.953 lg | 274.5 | 1.118 | 0.340 | 0.078 |
| 14 | PMH | | 0.41 | 10 500 | 39.7 | 4.0 | 48 | −2.95 | 2.70 | −25.01 | 29.02 | 38.31 | 40.75 | 0.260 | 0.993 lg | 286.4 | 1.070 | 0.284 | 0.065 |

*See text for discussion
PMH = PM/reluctance hybrid
REL = synchronous reluctance
SPM = surface PM motor
lg = lagging
ld = leading
50°C temperature rise assumed
Volts are a.c. r.m.s.